THE STEPHEN BECHTEL FUND 

IMPRINT IN ECOLOGY AND THE ENVIRONMENT

The Stephen Bechtel Fund has established this imprint to promote understanding and conservation of our natural environment.

The publisher gratefully acknowledges the generous contribution
to this book provided by the Stephen Bechtel Fund.

Laws, Theories, and Patterns in Ecology

Laws, Theories, and Patterns in Ecology

Walter K. Dodds

UNIVERSITY OF CALIFORNIA PRESS

Berkeley *Los Angeles* *London*

University of California Press, one of the most distinguished university presses in the United States, enriches lives around the world by advancing scholarship in the humanities, social sciences, and natural sciences. Its activities are supported by the UC Press Foundation and by philanthropic contributions from individuals and institutions. For more information, visit www.ucpress.edu.

University of California Press
Berkeley and Los Angeles, California

University of California Press, Ltd.
London, England

Library of Congress Cataloging-in-Publication Data

Dodds, Walter Kennedy, 1958-
Laws, theories, and patterns in ecology / Walter K. Dodds.
p. cm.
Includes bibliographical references and index.
ISBN 978-0-520-26040-5 (case : alk. paper)—ISBN 978-0-520-26041-2 (pbk. : alk. paper) 1. Biotic communities. 2. Ecology—Philosophy. I. Title.
QH541.D64 2009
577—dc22

2009010052

Manufactured in The United States of America
16 15 14 13 12 11 10 09
10 9 8 7 6 5 4 3 2 1

The paper used in this publication meets the minimum requirements of ANSI/NISO Z39.48-1992 (R 1997) (*Permanence of Paper*).∞

For the Ecologists

CONTENTS

PREFACE

This book is an attempt at first principles of ecological systems. I still can recall, as a young ecologist, my first real thinking in general terms about how all species interact. I was walking along the banks of the Willamette River in Oregon contemplating a "tangled bank," as Darwin had mused over dense vegetation more than 100 years before. I was thinking about how intense the competition must be, and then realized the species on the riverbank could be facilitating each other and that competition was my assumption from my ecology courses so far. This contemplation started my general questions about assumptions of ecology and the journey of curiosity-driven general ecological research I have been fortunate enough to take.

The search for general principles (i.e., my own set of assumptions and framework of understanding) started in 1993 with an end-of-semester walk to the corner pub with my longtime colleague and good friend, John Blair, to celebrate the end of classes. We discussed the idea of first principles over a pitcher or two, and

I played with the idea only occasionally after that time. A few years ago, I started seriously thinking about the possibility of simplifying principles of ecology and in quick succession made a seminar for the ecology and evolutionary biology group at Kansas State University and an extended version for Jim Brown's biocomplexity group at the University of New Mexico. Both meetings were very stimulating; my 50-minute presentation was followed by over an hour of argument. Discussion at the New Mexico meeting was particularly spirited. After the argument with the biocomplexity group, a student commented on how intensely the group criticized my ideas; I was more intellectually stimulated by this meeting than most any other in my academic life. This topic has since led to numerous vigorous discussions on the general topic. If nothing else, I hope to transmit the intellectual excitement of those discussions to readers of this book.

This book is intended as a graduate-level point of departure for individual thought or group discussion. The book is simple enough that it could be used by advanced undergraduates, but I assume a general ecological knowledge and do not explain general ecological concepts except as they relate to a system of first principles or how they present difficulties in prediction. Eugene Odum (1959) took a similar organizational approach to first principles when he wrote his influential text, *Fundamentals of Ecology*. Reportedly, he wrote it as a response to biologists in his department who thought that ecology was a field with no rigor or basis in theory. They viewed ecology as nothing but natural history (the role of natural history in ecology is discussed in Chapter 2). As this book took shape, I was amazed how many predictions could be made by combinations of first principles and how far ecology has come given the relatively young state of the science. Now we are being called to solve one of humanity's great challenges—finding a way to live on Earth without destroying the capacity of ecosystems to support us. My hope is that young scientists will turn a critical eye toward their research and that of others with respect to its contribution to a predictive science.

Murray (2000) proposed that one of the reasons a general system of ecological laws has not been put forth is an unwillingness to be called wrong. He then proposed ecological laws as an inducement for attempts to falsify them. At least the approach adopted by Murray and in this book could potentially start the Popperian ball rolling (i.e., the laws proposed in this book should be treated as hypotheses to be tested). Ecological science, as all other science, is a field of changing paradigms (Cuddington and Beisner 2005) and my approach here is likely just a snapshot of changing ecological thought at one time. My attempt to build a framework of laws is, at a minimum, an exercise to organize my own thoughts, and I can only hope this exercise is of some use to others. The ideas presented could provide a fruitful topic for discussion among ecologists. Maybe this book can serve as a study guide for doctoral preliminary examination preparation. Be careful if it is used this way because my point of view is not always in the mainstream. I particularly anticipate the possibility that graduate students will think deeply about the issues presented and look forward to finding out from them how spectacularly wrong the ideas presented in this book are. If none of my laws have value, my alternative list of laws is short:

1. The only absolute law in ecology is that there are no laws.
2. Statistics are better than obvious results.
3. All predators are voracious.

4. The conservation value of an organism is positively correlated with the relative size of its eyes and the degree of fuzziness (feathers or fur) of its young.

5. Ranges of obscure species tend to center on institutions employing systematists who specialize on those groups.

6. The depth of any shallow aquatic habitat is always a few centimeters greater than the height of the researcher's boots.

Just as life is too short to take everything seriously, it is also too short not to take some things seriously. My approach here is one of serious reductionism; reductionism will be addressed directly at the end of Chapter 1. In this book, it is assumed that first principles apply and can be used to make some predictions in complex systems. If reductionism fails, then that discovery is of value as well. The expansion in study of complex systems is a response to such reductionism, though the search for emergent properties is a search for first principles across complex systems.

Thanks to all readers for taking time to consider the points I have been mulling over all these years. Hopefully the ideas will fuel thought and discussion. Feel free to let me know what you think.

ACKNOWLEDGMENTS

I am grateful for comments on various portions and ideas for this book from J. David Allan, Sarah Baer, John Blair, James Brown, David Hartnett, Bob Holt, Dawn Kaufman, Don Kaufman, Bill Lindberg, Bruce Milne, Robert Rickleffs, C. C. Smith, Matt Whiles, Samantha Wisely, and Lydia Zeglin. Several of my colleagues at Kansas State University—Joe Craine, Keith Gido, Tony Joern, Jesse Nippert, and Brett Sandercock—supplied detailed helpful comments. Two anonymous reviewers provided helpful reviews of the book. Peter Turchin reviewed the book, improved my definition of laws and pointed out my working title was misleading. Many of these people did not agree with my points, but they helped clarify my thinking and kept me a little more honest (helped me curb my enthusiasm). Dolly Gudder proofread repeatedly and helped with references. My editor, Chuck Crumly, provided valuable input in both content and presentation. There are not too many science editors that show up for technical talks at conferences in areas outside their immediate expertise, but Chuck does, and the broad intellectual framework this supplies is evident in his input. Our discussions over sushi were quite valuable and are fondly remembered, even if his credit card did not always work as planned. Jamie Gillooly kindly provided raw data on metabolic scaling. Thanks to members of the University of New Mexico Biocomplexity Group and the Kansas State University Ecology and Evolutionary Biology seminar for helpful criticisms and suggestions. Discussion over these topics was also spurred by numerous contacts and discussions from seminars at Southern Illinois University Carbondale, Zoology, the University of Florida, Fisheries and Aquatic Sciences. I am grateful for invitations to present part of these ideas at those institutions. I am particularly thrilled by how excited graduate students got about these ideas; the future of our science is in good hands. I thank the Division of Biology at Kansas State University and the National Science Foundation, particularly the Konza LTER and Kansas EPSCoR Ecological Forecasting grants, for funding during the writing of this book. Thanks to Christopher Robinson and EAWAG for hosting us during proofing of the book. This is contribution no. 09-001-B from the Kansas Agricultural Experiment Station.

Introduction

> The empirical basis of objective science has thus nothing 'absolute' about it. Science does not rest on solid bedrock. The bold structure of its theories rises, as it were, above a swamp. It is like a building erected on piles. The piles are driven down from above into the swamp, but not down to a natural or 'given' base; and if we stop driving the piles deeper, it is not because we have reached firm ground. We simply stop when we are satisfied that the piles are firm enough to carry the structure, at least for the time being.
>
> *Sir Karl Popper 1968. The Logic of Scientific Discovery, p. 111*

PREDICTION IN ECOLOGY

All species need to predict features of their environment or they will not survive. Capacity for prediction could be a basic genetic program as a fixed response to environmental cues or prediction could be the capacity to learn from the environment. Successful forecasting of ecological properties is the hallmark of evolutionary success; matching environmental and ecological context to resource needs for reproduction and survival is a requirement.

Prediction is innate, and ecological prediction was required for hominid survival throughout most of our history. Humans (*Homo sapiens*) have attained the ability to predict the environment better than any other species (at least on the short term) and, subsequently, have become the most successful single species of herbivores, predators, and competitors, dominating most ecosystems on Earth.

The desire to predict is a deep-seated component of human nature, and scientists have formalized this desire to the greatest degree in human endeavors. The ultimate goal of many scientists is to be predictive at the broadest possible temporal and spatial scales. The importance of understanding to science has also been stressed (Pickett et al. 1994); understanding can help predict mechanisms as well as patterns. Prediction is generally viewed by scientists as more rigorous than the accommodation of explaining past patterns (Lipton 2005). Thus, prediction holds a special place in the goals of science.

Levins (1966) suggested that inherent trade-offs in prediction require that any one of three strategies to study natural systems can be sacrificed to pursue two others: (1) sacrifice generality to realism and precision, (2) sacrifice realism to generality and precision, or (3) sacrifice precision to realism and generality. If we assume realism is essential to ecological studies and the goals of prediction are to maximize accuracy and precision, that leaves the first and third strategies. If Levins is correct about trade-offs, general laws with predictive power should be difficult to come by. If laws provide predictability at the broadest scales, they will have high scientific value.

The more complex a system, the more difficult it is to find simple mechanisms controlling processes. Ecology is one of the most complex fields of science, and by extension, simple laws that can be used to make relevant predictions are more difficult to come by in ecology than in some other fields. Despite this difficulty, identification of laws as a foundation of ecology may be useful because formulation of basic principles can clarify the scientific basis for ecological theory and application.

Relevance of science to human needs is not necessarily directly related to predictive ability of a particular scientific field. Mathematics and physics are probably the strongest branches of science with regard to generality and accuracy of prediction, respectively. However, the specific relevance of advanced mathematics or physics to predictions that humans want to make is often indirect, and mathematics and physics of 200 years ago are adequate for many predictions we want to make. Conversely, the specific relevance of many current biological problems, human health, for example, is highly pertinent to humans. Though predictive ability is moderate for many health issues (e.g., how likely a person is to recover from a specific cancer), human investment in the science to allow even minor improvements in prediction related to health issues is substantial.

As with all biological sciences, the field of ecology needs to deal with complex systems. Ecology is the most complex biological science because it requires consideration of all levels of biological organization from genes to global ecosystems. Furthermore, there is a historical context (evolutionary and otherwise) for each ecosystem. There is an ongoing attempt by practitioners of ecology to untangle ecological complexity (e.g., Maurer 1999). The stakes are high because many issues in ecology—global climate regulation, production of natural resources, conservation of ecological services, and preservation of species—are of great interest to humanity (Dodds 2008).

Since its inception, the field of ecology has been home to debate over the existence of ecological laws and principles and possibilities for prediction, and the theme of laws in ecology has periodically reappeared in ecology literature. Particularly relevant to this book, a group of recent publications address the controversy of whether laws exist in ecology (e.g., Lawton 1999, Murray 2000, Turchin 2001, Colyvan and Ginzburg 2003, Lange 2005, O'Hara 2005, Scheiner and Willig 2007, Lockwood 2008).

In the early days of ecology, some scientists were unafraid of claiming generality, whereas others felt laws were difficult to support (McIntosh 1985). For example, rules erected around succession theory by some North American ecologists reflected the idea that predictable changes occurred in ecological communities that could be described by a series of laws or rules of assembly. European phytogeographers made similar attempts at generality. Theoretical ecology blossomed during the middle of the 20th century; numerous investigators published broad predictions that received substantial, uncritical support (e.g., diversity-stability arguments, island biogeography, species-area relationships, the role of competition in communities). One influential text that undoubtedly influenced the theoreticians, *Fundamentals of Ecology*, (Odum 1959) proposed a series of principles or statements on which ecology is built. Thus, in the early 1970s, ecology was a field full of promise with sweeping generalizations on the horizon and technology (computers, complex modeling, new sensing equipment from satellites to small-scale sensors) to explore this new territory and the terrain beyond.

As ecology matured as a science, criticisms of lack of experimental and statistical rigor decreased the degree of uncritical support for general theories. Lack of statistical rigor was a common problem in many studies (Hurlbert 1984). The role of competition in structuring ecological communities came under intense scrutiny. It was assumed by many scientists up to this time that competition was the primary factor structuring ecological communities. This assumption allowed scientists to construct assembly rules based on observed patterns. However, these rules were rarely based on direct experimental evidence, and more commonly on observations of existing patterns. Some scientists argued it cannot be assumed that competition occurs between any two organisms without rigorous experimental testing of the relationship; others argued that competition is a central force structuring ecological communities (e.g., Connell 1983, Schoener 1983). An extension of the debate over the specific role of competition in structuring plant communities exemplifies the continuing debate over the role of competition and its ecological implications (Craine 2005, 2007). Such dialog strengthened the experimental side of ecology but did little to establish the comprehensive theoretical basis required for a scientific field of study. The debate probably caused some to shy away from proposing laws in ecology.

In one of the strongest attacks on the status quo of contemporary ecology, *A Critique for Ecology*, R.H. Peters (1991) claimed that a major failure in much current ecology is the lack of predictive ability. He classified many major ecological statements as (1) trivial and obvious, (2) truisms, (3) tautologies, or (4) useless because their exception rate is too high. Although this assessment may be excessively pessimistic, Peters' points merit close consideration. He claims that predictability is central to success in science and that ecologists, for the most part, have not clearly delineated a scientific approach that can provide reproducible and general prediction. It is easy to be pessimistic about the inability to find general patterns and predict ecological responses given the variance and complexity inherent in ecological systems. Pickett et al. (1994) make a good case that progress will be much slower if empirical patterns are the only method for scientific prediction, and that creation of disciplines and subdisciplines and subsequent integration moves the field forward. Exploring the potential for laws occurring in the science of ecology can be useful for describing what we know; so much the better if these laws can be used to predict ecological phenomena and build ecological theories.

WHAT SHOULD WE EXPECT FROM LAWS?

Here, the possibility is explored that there are fundamental laws in ecology. Even though the field of ecology is as notorious as any other for semantic obfuscation, exploring the meaning of words and establishing a common language is a necessary first step to building a theoretical foundation. Webster's collegiate dictionary (1975) contains the following definitions of relevant terms that can serve as a starting point for discussion. The definitions have changed little over the last 30 years; the current Merriam-Webster online definitions are the same. At the end of this Introduction, I provide my definition of a law.

Law: a statement of an order or relation of phenomena that so far as is known is invariable under the given conditions

Principle: a comprehensive and fundamental law, doctrine, or assumption

Theory: a plausible or scientifically acceptable general principle or body of principles offered to explain phenomena

Theorem: an idea accepted or proposed as a demonstrable truth often as a part of a general theory

Proposition: a theorem or problem to be demonstrated or performed

Axiom: a proposition regarded as a self-evident truth

Postulate: a hypothesis advanced as an essential presupposition, condition, or premise of a train of reasoning

Mechanism: the fundamental physical or chemical processes involved in or responsible for an action, reaction, or other natural phenomenon (as organic evolution)

Tautology: a statement that is true by virtue of its logical form alone

Contingency: something liable to happen as an adjunct to something else

Constraint: the act of being forced by imposed stricture, restriction, or limitation

These definitions are but one way to view the relevant concepts. In this section, I discuss which of these definitions will be used and how they are modified in this book. Of the concepts conveyed in these definitions, I think *law* and *principle* best describe terms that form the basis for an underpinning of scientific study. The term law seems stronger to me than the term principle. O'Hara (2005) suggests principles are less rigorous than laws and, as opposed to laws, are sufficient for ecologists. Ecology should not settle for less rigor, but given the dictionary definitions listed previously, it is difficult for me to distinguish a large difference between principles and laws. The definition of principles includes not only laws but also doctrines and assumptions. Laws also have assumptions. I suggest for the sake of simplicity that law is as good a term as any to signify the basis of a scientific field, and this term is used to further examine some ideas about what constitutes a law.

Philosophers of science are generally unable to produce a commonly agreed on definition of what constitutes a law (Cooper 2003). Some even doubt the existence of laws at all (Giere 1999). The argument goes like this. A law must hold universally and there is no law that does that. Take for example the law of gravitation, which is one of the most widely accepted laws. For the law to be exactly tested, there must be two precisely spherical bodies under no influence from any outside force. Because there is no place in our universe where the gravitation of other bodies has no influence and no perfectly spherical objects, there is no place where the law of gravitation can be tested in its pure form. Currently, this view of no laws is in a minority, but it brings up the very important point of what degree of contingency or wiggle is acceptable in a law.

The debate will mostly be avoided in this book, but briefly covered here. I will use the general idea from Carroll (1994) that a law of nature is true, contingent, and general. It is important to recognize that some degree of contingency is acceptable in a law. The problem with contingency is that the degree of generality is limited, so the question of how general or universal a law needs to be to really qualify as a law remains open. Carroll (1994) recognizes potential problems with universality. With respect to laws in biology and ecology, we only know of life on our planet, so biological laws may well be universal, but we do not know for certain if universality is the case. The degree of acceptance of loss of generality to specific contingencies is subjective and thus what is, and what is not, a law is by definition subjective.

Various terminologies have been proposed to define the foundation of ecology. Turchin (2001) discusses some of these alternatives to the definition of laws in ecology, claiming that elementary propositions accepted without proof are postulates, and these can be used to derive theorems. Axioms have a definition similar to what Turchin calls elementary propositions. What Turchin (2001) defines as theorems and axioms are included in what I define as laws. The basis of each law (e.g., if it is axiomatic) will be discussed when each law is presented.

Brown et al. (2003, p. 410) suggested that a scientific law depends on the requirement that the mechanistic process and the conceptual framework behind the process are "universal or very nearly so." Such a requirement specifically discounts empirical patterns that are observed frequently but cannot be explained mechanistically. Mechanisms and axioms provide building blocks for laws or principles. Empirical patterns may have high statistical significance, but without mechanism, they remain mere observations. An observation such as the parabolic flight of a thrown ball is consistent and predictable, but it is not a law; laws of gravitation and motion lead to prediction of the predictable pattern.

We are on slippery ground if we require an absolute fundamental mechanism as a basis for a law (Ginzburg and Colyvan 2004). Although a few would argue the utility and predictive power of the law of gravity, this is also a good example of a law in which the fundamental mechanism is not understood.

> Isaac Newton recognized the law of gravity. But he never understood it. He never could explain it. Neither could Albert Einstein. Even today, neither can we. Still, all of us believe in gravity. None of us is ready to take a leap off the Eiffel Tower in hopes that gravity is merely a weak empirical generalization and not a law of the universe.
>
> *Rosenzweig (2003, p. 141)*

A law requires a mechanism, but ultimately, the mechanism does not have to be completely understood.

I assume (1) an ecological law is a statement of an order or relation of phenomena that so far as is known is invariable under the defined conditions and (2) the mechanistic process is known or that the law is axiomatic (self-evident). Mechanisms behind ecological laws can be taken from biological or other sciences but axioms or tautologies (self-evident truths such as mathematical identities and statements that must be true by logic alone) are also necessary to form many laws. My concept of laws does not include "probabilistic laws" (Pickett et al. 1994); the equivalent of "probabilistic laws" will be considered as strong patterns and are essential for combining with laws to build ecological theories as presented in Chapter 2. After more formally considering properties of laws, I will make a more explicit operational definition of laws.

It is useful to consider what is widely accepted as a scientific physical law to explore the features of what is and is not accepted as a law in science. The law of gravitation discussed previously is a good example for several reasons that are pertinent to discussion of laws in ecology: (1) it is accepted as a useful predictive law by a vast majority of scientists, and a simple equation can be written (relating the mass of two objects and the distance between them to the force of gravitation) to predict a physical characteristic of nature; (2) it holds precisely only under conditions that are not ever exactly met (e.g., absence of friction or absence of outside influences); (3) the actual forces that cause gravity are not completely understood (the law is axiomatic, it assumes that there is some force called gravity without proving the mechanism); and (4) some simple problems that are clearly governed by the law (e.g., the three-body problem of how three masses interact with each other gravitationally) cannot be solved by equations generated with the law.

Laws can have more than one mechanism. Mayr (2004) argues there are not laws in biology, only concepts. He suggests this is because there are two causes for everything that organisms do: (1) the genetic program all organisms abide by that is formed by evolution and (2) natural laws. However, physical laws have more than one cause or deal with several driving variables. For example, in the ideal gas law, $PV = nRT$, where P = pressure, V = volume, n = number of moles, R is the universal gas constant, and T = temperature, P will increase if V decreases, n increases (there is a higher density of molecules), or T increases (the average velocity of the molecules increases). There are multiple mechanisms that could cause an increase in pressure, but the law remains predictive.

Ecological laws could differ from laws of physics or other fields, making ecology autonomous (Lange 2005). Some proposed ecological laws have little to do with physics, but it may be difficult to clearly delineate laws that apply to only one scientific field. For example, there is considerable overlap between laws of physics and chemistry, even though the two are considered distinct fields. The ideal gas law is presented in introductory texts of both physics and chemistry. Fundamental biological laws that set the field of biology apart from physical and chemical sciences stem from laws of natural selection and modification with descent and specific constraints of organisms (e.g., the requirement for DNA).

O'Hara (2005) suggested that ecological laws are not valid if they are axiomatic (i.e., mathematical theorems or logical truisms) because they are not falsifiable, so they are not useful. Popper (1968) also made the same point with respect to all science, that axiomatic statements are not testable because they are not falsifiable.

Adherence to this Popperian approach could be an unrealistic limitation to scientific progress (Pickett et al. 1994). Axioms can form important parts of null laws. Also, axiomatic laws, such as mathematical identities, are useful foundations and, in my opinion, can be considered laws.

Perfectly good mathematical constructs (e.g., negative populations) can violate ecological or physical laws, but logically or mathematically based laws are useful within the constraints of realistic assumptions. Axiomatic statements allow clear delineation of the full range of possibilities and require that more restrictive statements be tested. Philosophers of science now consider Popper's criteria too restrictive and not reflective of the way that all scientific progress is made (Pickett et al. 1994). Consider a potential axiomatic law that applies to ecological communities: Species can have positive, negative, or neutral direct interactions (an early observation of ecologists, such as Odum [1959]). This law is axiomatic (there are no other logical possibilities), but it has some use. If ecologists assume otherwise, such as when "competition communities" are modeled, they should recognize they are making an assumption about the lack of specific kinds of interactions that may not be generally true in all ecological communities. Making this axiomatic statement (of all signs of species interactions being possible) clear requires any ecologists who think, for example, that competition and predation are the dominant forces constraining community structure to eliminate other logically possible interactions from consideration before their view offers true predictive ability.

In my treatment here, laws from other fields are considered axiomatic. The law that matter can be neither created nor destroyed can be considered axiomatic. Ecologists do not test this law, they take it as a given. Conservation of matter can be tested because reactants and products of chemical reactions can be measured and accounted for. Applying this law to ecological systems requires the axiomatic assumption that there are not negative concentrations of chemicals (unless you allow for antimatter, but that allowance is not ecologically relevant). We have no way to test for negative concentrations of chemicals, so conservation of matter applies to chemical reactions and biological processes. Theory of ecological stoichiometry as well as major parts of ecosystem theory (as approached in Chapter 2) have conservation of matter as a solid underpinning because the contingencies (antimatter, thermonuclear reactions) of this law are not too restrictive.

A law can be too restrictive to be useful if the defined conditions (constraints or contingencies) are too limiting. It seems that constraint and contingency are used interchangeably in the literature. However, given the definitions of both, constraint seems more absolute, so the term constraint is used when defining the conditions under which laws and theories hold. Lawton (1999, p. 177) argues that ecology has numerous laws (defined by him as "a generalized formulation based on a series of events or processes observed to recur regularly under certain conditions: a widely observable tendency"), but few of them are universally true. As constraints narrow the conditions under which laws are expected to operate, the more likely it is a law will hold absolutely. This line of argument opens up the question of how restrictive conditions should be before a law has narrow predictive ability and, therefore, limited use.

It is a mistake to assume laws have no exceptions (Colyvan and Ginzberg 2003, Turchin 2003, Ginzburg and Colyvan 2004). At the most fundamental level, Godel's theorem (in my admittedly naive understanding) states the impossibility of defining a complete set of axioms that do not give rise to contradictions in higher order systems. Laws used to make highly accurate predictions clearly have exceptions and constraints. Newtonian laws do not hold at subatomic scales. No perfectly elastic collisions or completely frictionless systems exist. It is also possible that all gas molecules in a closed container will hit one side of the container at once. In this case, there would be a catastrophic and rapid increase in pressure on one side of the container and a collapse on the other sides (i.e., a directional explosion). Such explosions are not observed because the probability that all molecules will hit the edge of the container simultaneously is so small. So, ideal gas laws still work. Constraints are always present, and ability to use laws in a predictive fashion depends on ability to define them. For example, the law of gravitation is not very good at predicting the rate of descent of a parachute. If a term for drag is introduced along with the effects of gravitation, a reasonable prediction of descent rate can be made. But, drag is a function of turbulence, updrafts, downdrafts, and temperature. Turbulence is a notoriously difficult problem in physics. Parachutes work, usually, but without precise prediction of exactly where and how fast the object attached under them will impact the earth.

If the exception rate is too high or conditions are too stringent for a law to hold, predictive value is low. Ecology is not a successful science if it does not allow for predictions to be made. Poff (1997) claimed that a fundamental goal of basic science is to understand how pattern is generated by process and the goal of applied science is to predict without requiring mechanistic understanding, but it might not be possible to reliably predict ecological patterns and processes if they are not truly understood (i.e., prediction is superior to accommodation). As systems are considered where conditions fall outside previously studied conditions, mechanistic prediction is more likely to work. Thus, I ask in this book if there are useful general laws (laws that offer predictive ability) and if reliable theories can be built from those laws in ecology.

Constraints and exceptions are a part of all laws. Cooper (2003) proposes the term law be abandoned in favor of *nomic force*. He goes on to suggest that nomic force can have various degrees of wiggle in explanatory power. It is unlikely that scientists will adopt his terminology, but it is useful to think about the degree of contingency associated with a predictive statement. The laws presented here have constraints, and the amount of constraint that can be tolerated and still allow me to consider a statement a viable law is subjective and difficult to quantify.

HOW LAWS WILL BE APPROACHED

My general approach is borrowed in part from a more specific route taken by Sterner and Elser (2002) in their monograph on ecological stoichiometry. They take simple initial propositions or laws (conservation of matter, shared biochemical composition of organisms) and build an entire body of theory on them. The fundamental strength of this process of simplification, or abstraction, is that it defines essential properties of interest and allows ignoring contingencies (Pickett at al. 1994). Jørgensen and Svirezhev (2004) use a similar approach of building from basic principles. They start with thermodynamic and chemical principles and build a theory of ecology as defined by these principles (and, as discussed here, perhaps take these ideas a bit far). I hope to emulate and broaden the general approaches of building from simple concepts taken by these authors.

Wilkinson (2003) took an even broader approach. He considered characteristics that would be necessary to sustain life on other planets over the time span of biogeochemical cycling and conducted a thought experiment (an experiment that is definitely more popular in physics than in ecology) that did not limit properties of ecology to those found on Earth. This is a useful exercise for exploring some basic laws of ecology, but the approach is too general for my purposes because it does not take advantage of specific properties of life on Earth, which can be used as constraints that improve predictability.

The approach of starting with simple laws as building blocks is just one of many general approaches to the study of ecology. Another approach is to rely mainly on pattern as the best predictor of ecological characteristics. Peters (1991) advocates this approach to ecological inquiry. This statistical view is supported by the argument that small-scale approaches are unlikely to reveal large-scale patterns because ecologically relevant patterns depend on large-scale processes. Thus, considering ecological patterns at large scales (i.e., macroecology) allows for a large enough sample to statistically resolve ecological pattern (Maurer 1999). The pattern approach can be taken further by assuming that if a strong pattern holds across systems and orders of magnitude, there must be a fundamental mechanism underlying it. The pattern approach is taken by some scientists looking for prediction of macroecological phenomena; patterns are used as clues to find applicable mechanisms (Brown et al. 2003). A third possible approach is to assume that each ecological system has its own contingencies or idiosyncrasies and that comprehending all local constraints is the key to ecological understanding. Simberloff (2004, p. 787) discusses this approach: "Laws and models in community ecology are highly contingent, and their domain is usually very local." Scheiner and Willig (2005) took the approach of defining contingencies; they proposed broad theories that could be used to develop models if local contingencies were taken into account.

My caricature of approaches to laws in ecology is crude but hopefully illustrates how separate approaches can be used in the same science. Some may find my approach too simple; a similar approach has been used as the basis for some introductory textbooks (Odum 1959, Krebs 1988). Still, I am surprised how far a simplistic approach can go toward explaining ecological complexity and how this approach can help clarify thinking about ecology. This book represents just one of several potential approaches. Logical consequences of defining fundamental laws at the outset and building theories from those laws are explored. I claim neither originality nor a synthesis that will solve major ecological problems. Ecology is flourishing as a science without the laws as proposed here, and the "laws" proposed here could be so obvious there is no need to elucidate. My hope is that such laws will slightly solidify the foundations of our field (or, in Popper's words, drive the piles that support the theory a bit deeper).

This framework of ecological laws can be viewed as an exploration of the unarticulated processes I would go through if exposed to a new ecological system; what would be expected in the absence of any specific information about the particular ecosystem? An extreme example of deciding how to decipher properties of a new ecosystem from the field of ecology is the discovery of dense aggregations of organisms at the periphery of deep-sea hydrothermal vents. Biologists knew there must be an energy source feeding these organisms in what is an otherwise oligotrophic environment and, subsequently, found the sulfur-oxidizing bacterial endosymbionts associated with the rift tubeworm *Riftia* (Lutz and Kennish 1993). Prior expectations from studies of other ecosystems are the existence of consumers subsisting on producers and potential competition among these consumers. Also, these organisms probably have the ability to disperse to other similar habitats or evolved in specific locations. These underlying laws or principles were probably unarticulated but shaped the approach researchers used to examine this unique habitat.

My framework contains 35 proposed laws. Some might object that this is too many to form the basis of a field of science. Scheiner and Willig (2007) propose only seven laws to form the basis of a general theory of ecology. Versions of all of the laws they propose, and more, are here. It is a mistake to think that only a few laws should be necessary to explain ecological systems (Rosenzweig 1995) or any other science. The field of physics has numerous laws such as Newton's laws of motion and gravitation, conservation of energy, laws of conservation of charge and mass in chemical reactions, Boyle's law, Charles' law, the ideal gas law, Hooke's law, the laws of thermodynamics, Coulomb's law, Gauss' law, Faraday's Induction law, the laws of optics, and others. Physics is still considered a rigorous science even though (or perhaps because) it has many laws.

Hopefully, the laws proposed here apply to entire ecological systems. My background is strongly microbial, and it leads me to generality that is not present in some treatments of ecology. For example, some laws have been proposed that apply strictly to animals, particularly those that require sexual reproduction, or higher plants. These laws may be useful within their constraints, but the laws are considered too specific for my current treatment because I think a law is too constrained if it does not apply to two major domains of life (Bacteria and Archaea) that make up the majority of protoplasm on Earth. My hope is that a general conceptual framework of ecology will apply to all organisms. However, just as the ideal gas law is not relevant to all chemical problems, each law does not need to apply to all ecological problems.

It probably is not possible to test the general predictive ability of many of the laws proposed here. The ultimate predictability would be to discover life on another planet and observe whether it follows the proposed laws after relaxing contingencies particular to our planet (Wilkinson 2003). There also may be some highly isolated ecosystems left in deep subterranean habitats that could be used to test existence of the proposed laws. But even if all the laws cannot be tested, observation of existing pattern can be used to bolster scientific conceptual frameworks. Other scientific fields such as astronomy, geology, and archeology are based heavily on observation and how a body of laws can be used to predict known observations. Lack of possibility of replication (n = 1 biosphere) and ultimate prediction of a completely new event does not mean laws that can form the basis of science do not exist.

In Chapter 1, laws are proposed and some other previously proposed laws are discarded. Then, I attempt to build ecological theories from the proposed laws. This approach was mentioned by Scheiner and Willig (2005), who stated that unified theories can be built from individual models. Theories have their own philosophical problems (Stanford 2006), which are discussed briefly at the beginning of Chapter 2. In Chapter 3, I discuss some patterns or concepts that are commonly seen in ecological systems but are difficult to apply as laws or theories because of their limited predictive ability. I am not arguing that patterns are not important to ecologists, rather they should be looked for in ecological systems and may be applied to specific systems once the fundamental properties that make that ecosystem unique are known.

For my definition of general laws in ecology, the following is used: a mechanistically based or self-evident statement of an order or relation of ecological phenomena that so far as is known holds generally under ecologically relevant constraints. By invariable I mean: (1) the law allows prediction of a statistically significant pattern, regardless of what habitat or taxonomic group it is applied to, (2) the law is axiomatic, or (3) the law is taken from other physical sciences so is treated as axiomatic. Some researchers suggest not mixing axioms with laws (e.g., Turchin 2003), but in this treatment I do because axiomatic laws have value, and separating axiomatic and mechanistic laws would add complexity to my scheme of presentation. The axiomatic laws are identified as such when they are presented in Chapter 1. By ecologically relevant constraints, I mean at the temporal and spatial scales under which organisms operate and the constraint of taxonomic specificity is disallowed (e.g., a law that applies only to sexually reproducing animals does not apply to plants or bacteria so is not considered a general law here). This is not to say that more restrictive ecological laws that apply to smaller geographic areas or restricted subsets of organisms do not exist or are not useful. More restrictive laws are potentially too numerous to be considered in detail in this book.

CHAPTER 1

Laws

In this chapter, laws are organized roughly so that those pertaining to subsequent laws are presented first. The first approach taken is deductive, but an inductive approach is unavoidable for some of the laws. Laws are also grouped generally by topic and level of biological organization. Some of the laws lead to specific equations that can be used to describe relationships; others are simple concepts not profitably captured in equations. Some may seem self-evident (axiomatic), but explicit definition at the beginning is crucial to explaining subsequent laws and theories that depend on them. Some are null model laws; they state the simplest assumptions that must be shown not to hold if the simplest explanation is not used. In all cases, I attempt to list axioms and mechanisms for each law. In some cases, the mechanisms are previously stated laws. I define each law with respect to the scales across which it applies and also attempt to list constraints. For example, some laws apply across temporal and spatial scales, others across biological levels of hierarchy. Of course, these two are interrelated, but for the most part, they are considered separately when classifying laws. First, general foundations are proposed, then laws of biology and ecology. After proposing the laws, some candidates for laws are explored that either do not have enough support to be considered a law or offer limited general predictive ability (e.g., are more constrained than my definition allows). Finally, the potential for a reductive approach in ecology is discussed as well as how the proposed laws relate to reductionism.

FOUNDATIONS

1. Laws from Physics, Chemistry, and Mathematics

All physical and chemical laws hold as fundamental laws. The logic of mathematical relationships holds given biologically realistic constraints. Although there are many mechanisms for these laws, in the current treatment it is generally most convenient to assume well-established laws from physics and chemistry hold without discussing mechanism. I also assume that mathematical operations and identities apply without requiring proof (that they are axiomatic in the context of this treatment). Physical and chemical laws cover the abiotic template that constrains evolution, physiology, species interactions, material flux and conversion, and nearly all aspects of behavior and life history of most organisms. The applicable laws range widely; some relevant laws include gravity, properties of fluids, aerodynamics, and diffusion constraints. Laws of thermodynamics and balance of chemical equations have strong biological relevance and are treated separately. Properties of water and organic chemistry/biochemistry also are particularly important. Mathematics apply to ecological systems, but biologically relevant constraints need to be applied. Mathematical logic must be tempered with reality when applied to the real world. For example, Newtonian mechanics work mathematically regardless if time moves forward or backward, but application of these mechanical laws assumes that time moves forward (Kauffman 2008). Negative concentrations of chemicals or negative population numbers do not violate mathematics, but they do not occur in the real world. Laws of physics and chemistry have exceptions, but can provide a high degree of predictability within the arena in which ecology operates. Many physiological adaptations can be explained by the biochemical constraints of organisms (Hochachka and Somero 1984). For example, that matter cannot be created or destroyed (the basic law behind stoichiometry) is not strictly true given the existence of thermonuclear or matter/ antimatter reactions, but for organisms other than *Homo sapiens*, this is a law. Application of the law of conservation of matter to chemistry of organisms interacting with their environment has led to an entire subdiscipline, ecological stoichiometry (Sterner and Elser 2002). This field uses the proportion of elements in individual organisms related to resources that are available to them to make predictions regarding limiting factors. Thermodynamics and chemical interaction kinetics can provide rate laws for microbial respiration and growth relationships (Jin and Bethke 2003). Laws from physics and chemistry apply across all temporal and spatial scales; specific laws are applicable to more limited scales.

2. Evolution and Natural Selection

Natural selection will act on species leading to changes in genotype and, thus, phenotype. The mechanism is natural selection on heritable characteristics coupled with slight variability in the heritable characteristics. This theory is based on the Malthusian law of exponential growth in the absence of controls on growth, heritable characteristics that influence survival, and the selective process related to differential survival. The law of natural selection as a driver of change has been referred to as the first law of evolution (Murray 2000).

Mayr (2004) argues that Darwin's theory of evolution is actually composed of five theories and lists five subtheories of evolutionary theory: (1) the world is steadily changing, in part directionally (although not with an ultimate goal), and organisms are being transformed over time; (2) every group of organisms descended from ancestral species (common descent); (3) evolution proceeds gradually and never in jumps; (4) evolution explains why there are so many species; and (5) natural selection is a primary mechanism for evolutionary change. Of Mayr's five "theories," the criteria used in this book could be interpreted to mean that 1, 2, and 5 are actually laws. The third, that saltation or very rapid evolution is not important, is really a pattern because some exceptions occur. The debate over punctuated equilibrium has been vigorous. Examples of evolutionary jumps include endosymbiotic origins of organelles and stabilized hybrids that can reproduce without crossing (e.g., allotetraploids). The fourth, that evolution explains why there are so many species, could be thought of as a consequence of the first, second, and fifth that provides support for the general theory of evolution.

There are other laws of evolution. For example, Pickett et al. (1994) suggest that the Hardy-Weinberg principle is a "zero force" law in evolutionary theory similar to Newton's first law of motion. This principle is an example of an axiomatic law that provides a point of reference against which to compare other observations. They also note that kin selection and sexual selection can act in addition to "natural selection" (i.e., selection can be interspecific or intraspecific). The units of selection (genes, organisms, group selection) remain controversial.

Because natural selection is acting on variation that arises by chance, evolution throws a wrench in predictability of many other proposed laws (Lawton 1999, Solé and Bascompte 2006). Organisms are constrained by their evolutionary history. Mayr (2004) notes that evolution guarantees there are two mechanisms for all biological phenomena, evolution, and the immediate factors (biotic and abiotic) leading to the phenomena. Theory of evolution is different from the basic law or laws of evolution. Theory of evolution is what most of biology is concerned with. However, changes in characteristics with natural selection can be considered a law because not only biological systems evolve. If systems are set up with reproductive units that are allowed to reproduce slightly less than perfectly and those imperfections are heritable (e.g., computer programs that improve themselves, Koza 1991, Yedid and Bell 2002), then the process of evolution occurs.

Evolution applies from microbial to global scales. The theory leads to many predictions including the greater diversity with age of habitat (e.g., the animal diversity of old lakes such as Baikal and Tanganyika), adaptation to toxins and chemical defenses, predator/prey coevolution, and many others. Many of these predictions—antibiotic resistance, resistance of pests to pesticides, and evolution of microbes to bioremediate organic pollutants—are practical.

The theoretical basis of evolutionary theory is so broad that it is outside the scope of this book. Few modern scientists, and even fewer biologists, would dispute the validity and application of the theory of evolution. Alternatively, ecological principles are applicable to selective pressures that drive evolution, so laws and theories considered in greater detail in this book have relevance to evolutionary theory. Another way of looking at the relationship between the theory of evolution and ecological laws is that any broad pattern in ecology predictable by us should be predictable (and exploitable) by organisms over evolutionary time. Thus, evolution provides the basis for ecology, or at least is a strong pattern driving it.

3. *Dominance of* Homo sapiens

Humans interact with more species more strongly than any other species (we are the ultimate keystone species, Kareiva et al. 2007). The mechanisms are that cognitive ability and cultural evolution have led to technological abilities that allow humans to dominate the earth. Another way to view this success is that technology may allow us to be "jacks of many trades." Humans are still not as good at some activities as other species. We cannot control some pests without biocontrol agents. We co-opt other organisms for our uses such as crops for food production and microbes for sewage treatment and bioremediation because they are better than synthetic technological approaches. Still, we are capable of more than any other individual species. Human dominance of the biosphere is clear; one-third to one-half of Earth's terrestrial surface has been transformed, CO_2 in the atmosphere has increased by about one-third since the industrial revolution, half the planet's freshwater is co-opted for human uses, half or more of biologically available nitrogen (N) and phosphorus (P) input into the biosphere is human-derived, and about one-fourth of bird species are extinct as a result of human activities (Vitousek et al. 1997). Cyanobacteria were one of the great global transformers when they oxygenated Earth's atmosphere, but it is unlikely that a single species was responsible. Some ubiquitous species of bacteria may currently have global effects, but dominance by any single species seems increasingly unlikely given molecular data on bacterial diversity across the globe. The study of ecology in the current natural world should consider the potential influence of human activities when results are scaled to ecosystems. This foundational concept applies from microbial to global scales.

FUNDAMENTAL BIOLOGICAL LAWS

4. Biological Composition

All organisms are constructed with a common set of molecules. All organisms are composed of cells. Each organism needs a cell membrane (lipid bilayer), proteins, DNA, RNA, and other biological molecules. Exact proportions vary by organism and life cycle. For purposes of this treatment, the general composition of cells is axiomatic. Viruses are not considered organisms here. Discovery of the common biochemical basis of all life on Earth and mechanisms of homeostasis and heredity forms one of the great triumphs and foundations of modern biology. A constraint is that some organisms have requirements that others do not. For example, diatoms require a substantial amount of silicon to build cell walls, but most other organisms have a very low silicon requirement. This law applies at the scales of cells and organisms.

5. System Openness

There are no closed ecological systems (from organisms to ecosystems) with regard to energy or nutrients. Nutrients are defined as elements required for organisms to build cells. Organic carbon is a nutrient, but many organisms also require it as an energy source. Ammonia can be oxidized to nitrate by bacteria to yield energy. In this context, ammonia is not a nutrient; it is a source of energy. Still, the same bacteria require ammonia to build proteins so they can also use it as a nutrient. The mechanism for openness is that abiotic (and to some degree biotic) factors move materials and energy into and out of all areas of the biosphere. Two equations can be used to account for energy and nutrient flux, respectively.

$$E_{input} = E_w + E_s + E_{output} \tag{1}$$

where E_{input} = energy input to the system, E_w = energy dissipation by biota due to the fact that physiology is not 100% efficient, E_s = energy storage, and E_{output} = energy transported out of the system.

$$V_{input} = V_{storage} + V_{output} \tag{2}$$

where V_{input} = the rate of nutrient input to a system, $V_{storage}$ = rate of storage, and V_{output} = loss of nutrients from an area. This law states that E_{input}, E_{output}, V_{input}, and V_{output} all must be greater than zero. A constraint of this law is the inability to exactly predict E_w and how much greater than zero E_{input}, E_{output}, V_{input}, and V_{output} all must be. System openness applies across all temporal and spatial scales.

6. *Recycling Rates*

For ecological systems in a prescribed area, energy will be depleted more quickly than nutrients without external inputs. This rapid depletion of energy is dictated by the laws that matter cannot be created or destroyed under normal conditions and that thermodynamics dictate the impossibility of complete efficiency of energy transformation and storage. This law functionally states that in the equations from the previous law $E_w > 0$ and that there is no term equivalent to E_w in the accounting equation for nutrient flux. A constraint of this law is the inability to exactly predict E_w. This law and the previous law form the fundamental basis of much of ecosystem science and theory of nutrient cycling. This law is scale independent.

PHYSIOLOGICAL CONSTRAINTS OF ORGANISMS

7. All Organisms Die

That all organisms must eventually die (or cease to exist as a single organism in the case of binary fission or vegetative reproduction) is an axiom because there is not a proven reason why a single organism cannot survive indefinitely. Even in multicellular clonal organisms, the original organism does not exist indefinitely. Scheiner and Willig (2007) propose a similar law (first principle) and claim correctly that mortality is less restrictive than senescence. Still, lines between death of an organism, continuation of a genome, and clonal reproduction make the idea of death a bit fuzzy. An organism that reproduces and evolves could eventually outcompete one that simply exists, thus natural selection would be an underlying mechanism for the idea that all organisms die and must reproduce to avoid extinction. The corollary to this law is that all organisms must reproduce. This law operates at the individual organism scale.

8. Energy Requirement

All organisms require energy for maintenance and reproduction. The first mechanism is that organisms are subject to laws of thermodynamics and that energy is required for homeostasis. The second mechanism is that growth, movement, and reproduction are biological requirements for continued survival. There is a requirement for reproduction because all organisms die, and organisms that do not reproduce will not have populations that successfully continue to evolve. Applying the concept of conservation of energy, this law can be represented by the following equation.

$$E_{intake} = E_g + E_r + E_m + E_s + E_w \quad (3)$$

where E is energy, *intake* = total energy intake, g = growth, r = reproduction, m = metabolism, s = storage, and w = waste. This law is similar to what Wilkinson (2006) views as the central concept of ecology, that organisms take energy from their environment and release waste back into the environment. Models of this general form, in which accounting for total energy allows for estimates of production, are based on this law and the law of conservation of energy from physics. Such models have been developed and applied by fisheries ecologists for some time (Wooton 1990). This type of model has direct applicability to ecological concepts. A constraint of this law is that the relative amount of energy per unit biomass varies across organisms and cannot be exactly predicted. This law operates at the individual organism scale.

9. Nutrient Cycling Requirement

Organisms must transform nutrients to operate. The first mechanism is biochemical; the molecular machinery of cells must be continuously renewed to regulate activity, repair damaged molecules, and allow for reproduction and growth. A second mechanism is that maintenance of homeostatic state (oxidation reduction potential, ions, pH) requires transformation of some nutrients. The law of system openness applies to this law. All biological membranes are somewhat leaky, and thus not perfectly efficient. This leakage dictates that nutrients cannot be conserved inside a cell indefinitely, so there must be uptake and loss of nutrients. Excretion of some nutrients is necessary to maintain homeostasis and allow survival until reproduction. Applying conservation of matter allows for accounting of rates of material flux in and out of organisms.

$$V_{internal} = V_{uptake} - V_{loss} \tag{4}$$

where V is the rate of change of nutrient per unit time, *internal* = nutrient concentration in the organism, *uptake* = acquisition from the environment, and *loss* = leakage or excretion from the organism to the environment. This law states that $V_{loss} > 0$, so over time, V_{uptake} must be at least equal to V_{loss} if the organism is to survive and must exceed V_{loss} if the organism is to grow and reproduce. The predictive constraint is that it is difficult to predict by how much V_{uptake} must exceed V_{loss}, and both can vary independently with time. Furthermore, relevant time scales vary. A bacterium may be able to survive only a few days with net nutrient loss, but a large tree may survive several years while losing nutrients. This law operates at the individual organism scale.

10. Maximum Metabolic Rates

Rates of individual metabolic processes exhibit saturation at high rates of nutrient resource supply. This saturation is a consequence of basic characteristics of enzymes. Enzymes operate under

Henri-Michaelis-Menten kinetics as represented by equations derived from first principles in chemistry,

$$V = V_{max}\, n/(K_n + n) \tag{5}$$

where V = rate of activity, V_{max} = maximal rate, n is substrate (nutrient resource) concentration, and K_n is half saturation constant, or the substrate concentration where $V = \frac{1}{2} V_{max}$. Because all biological processes are mediated by enzymes and only a finite number of enzyme molecules can be housed in any one cell, organisms must have a maximum sustained rate of uptake, processing, or growth. The constraint is that V_{max} and K_n vary within and across organisms and also as a function of abiotic conditions (e.g., temperature). This law operates from enzymes up to individual organism scale and can apply to assemblages of organisms.

11. Water Requirement

Organisms must have liquid water to grow and reproduce. Some organisms are able to withstand desiccation in an inactive state for decades or even centuries. However, organisms cannot continue metabolism when frozen solid or in boiling water. The mechanism for no growth and reproduction in solid ice is that nutrients cannot move quickly enough through ice to support growth. Water can be thought of as a vital nutrient, and if this is the case, the law of system openness states that leakage must be greater than zero, so organisms cannot survive without water indefinitely. The mechanism behind organisms not growing in boiling water is less defined but probably has to do with the instability of biological membranes and proteins at high temperatures. A constraint is that some organisms are considerably more efficient with water than others; some can withstand very high temperatures for a modest amount of time and grow very near the boiling point of water. This law operates at the individual organism scale.

12. Temperature Optimum

Each organism will exhibit a temperature optimum for growth and metabolism (i.e., a hump-shaped curve). The mechanism is the fundamental chemical properties of proteins, membranes, and other molecules. Diffusion rates increase with temperature increases. Chemical reaction rates that constrain metabolic activities increase exponentially with temperature according to the Van't Hoff-Arrhenius equation,

$$R = e^{-E/kT} \tag{6}$$

where R = rate, E is the activation energy, k is Boltzman's constant, and T is absolute temperature (degrees Kelvin). However, organisms must maintain structure, so metabolism does not continue to increase with increased T. The rate at which function falls off from the exponential is mainly fit by empirical means because the process of failure at temperatures above optimum is so complex. Some degree of failure could be occurring before optimum is reached, and the optimum could represent the sum of two conflicting processes, increase in diffusion and potential reaction rates with temperature and decrease in rate with excessive temperature as the molecular machinery of the cell begins to fail. Membranes must be fluid and able to maintain a stable bilayer to function properly. If the temperature is below that at which membranes can be fluid in a particular organism, the metabolic rate is zero (equation 6 does not apply). Protein activity increases with temperature until proteins begin to denature. However, at some temperature, function of biological molecules will be impaired. This law is one application of Shelford's "law of tolerance" (Odum 1959). Biological constraints in concert with evolution lead to specialization in temperature optima. A predictive constraint is the inability to predict the exact shape of the curve once failure starts to occur. This law operates from enzymes up to individual organism scale and may influence evolution rates (Allen et al. 2006) as well as physiological ecology of organisms.

BEHAVIOR OF ORGANISMS

13. Sensory Integration

Organisms with directed motility are expected to have the ability to sense changes in their environment over time. Sensory integration is a requirement for successful navigation. This law is axiomatic but requires laws of natural selection and recognition that motility has an energy cost, so movement without direction should be selected against except as a dispersal mechanism. Organisms that do not move directionally (e.g., plant seeds, passive larvae of some animals) do not need to sense changes in their environment. Because organisms move across a wide array of temporal and spatial scales, this law applies to individual organisms operating across those scales. The constraint is that these sensory systems have multiple evolutionary origins, so the specific type of sensory response cannot be predicted.

14. Predictability of Behavior

The best predictor of behavior of an organism or species is prior behavior of the same organism or species under the same conditions. The law is based in part on the genetic basis of many behaviors.

Predictability of behavior is also based on the concept that learned behavior increases survival; modes of behavior that were successful in the past are used again. Some unpredictability underlies this law because not all organisms have completely predictable behavior (unpredictable behavior, such as predator avoidance, can be an evolutionarily successful strategy). Not all individuals in a species will exhibit the same behavior. Much of behavioral ecology and even psychology, economics, and sociology are based on the concept of repeated behavioral patterns as applied to humans. Without this law, there would be no reason to conduct behavioral experiments. The major predictive constraint is related to the probabilistic nature of this law. The law applies from the scales of microbes to the largest animals.

FUNDAMENTAL PROPERTIES OF POPULATIONS

15. Conservation of Individuals

There is no spontaneous generation. The origin of life is the only logical exception to the law that must have occurred at least once. This law is axiomatic, though certainly has been tested often enough. The law leads to a fundamental equation used to budget population dynamics.

$$dN_{total}/dt = (dN_{birth}/dt - dN_{death}/dt) + (dN_{imm}/dt - dN_{em}/dt) \qquad (7)$$

where dN_{total}/dt = total rate of change of numbers in a population, dN_{birth}/dt = birth rate, dN_{death}/dt = death rate, dN_{imm}/dt = immigration rate, and dN_{em}/dt = emigration rate. $dN_{birth}/dt - dN_{death}/dt$ describes within patch dynamics, and the second half of the equation describes between patch dynamics. Turchin

(2003) proposes the form of this law without immigration and emigration as the first postulate for exponential population growth. This law applies at organism and population scales.

16. Exponential Growth

A population of organisms will increase exponentially (geometrically) given abundant nutrition and no disease, predation, emigration, or immigration. The constraint on this law is no controls on growth, and this has been taken to mean that it cannot be a law because this does not happen in a pure form (Lockwood 2008). Still, there are situations where constraints are removed temporarily. The mechanism for geometric growth is evolution. Any species consisting of individuals that do not grow and multiply have zero fitness. O'Hara (2005) makes the point that exponential growth is not the same as geometric growth. A continuous equation allows for fractional organisms, an unrealistic constraint with small populations, but is an assumption that makes the mathematics simpler (Rose 1987). However, if we assume the constraint of a large enough population and an appropriate time scale, an exponential relationship adequately captures an essential feature of change in populations upon relaxation of limiting factors. Applying numbers and scale is analogous to the ideal gas law; the gas law does not hold if the time frame is too short or if few gas molecules or atoms are considered at one time. The law of exponential growth can be stated as a foundational equation of population biology.

$$N_t = N_0 e^{rt} \tag{8}$$

where N_t is the population at time t, N_0 is the initial time, and r is the growth rate constant.

A more general statement of this idea is that if birth rates and death rates remain constant and emigration and immigration are equal (they cancel in equation 7), a population will change according to an exponential function (assuming birth and death rates are not equal). However, it is difficult to predict when birth or death rates will remain constant, so the more general statement offers less predictive ability (though as a fundamental mathematical law, exponential growth or decay explains numerous phenomena, Berryman 2003). This law applies at population scale.

17. Limits to Growth

No population can increase indefinitely. This law is axiomatic and a consequence of the fact that the world as well as local supply rates of resources are finite. Limits were pointed out by Aristotle 2,300 years ago (Krebs 1988). Most organisms can use only part of the world, inhabit restricted areas, and have defined resource use patterns. The specific factor that will limit population growth and size is a more difficult problem and is one of the primary predictive constraints with this law. Resource supply, competition (interspecific or intraspecific), or predation (including disease) can ultimately limit reproduction. Population density dependence is a common feature of species population time series (Brook and Bradshaw 2006). This law does not mean that the logistic curve will necessarily describe the population trajectory. There is no fundamental law or property that dictates the mathematical structure of the logistic curve, but the logistic is a simple mathematical way to include density dependence in a continuous equation (Rose 1987). Time lags and fluctuations in carrying capacity and growth rates are all possible and make it difficult to predict exactly when population level will be limited. Still, if rate of supply of a limiting resource is known, the final population that can be supported can be calculated. Our inability to calculate a human carrying capacity (Cohen 1995) exemplifies difficulties faced by ecologists in determining limits to growth. Inability of *Homo sapiens* to control their global environmental effects is symptomatic of their inability to comprehend the reality of this law. This law applies at the population scale. The self-thinning rule (or law) in plants is based on space-filling models of how much biomass can develop per unit area (Pretzsch 2002), and the limitation is photon flux density. Self-thinning is simply a function of efficiency of use of sunlight, and amount of sunlight per unit area dictates the population density that is possible. This efficiency varies among species, so the "rule" leads to somewhat variable scaling relationships (Pretzsch 2006).

18. Population Stability Not Determinant

The mathematical properties of density dependent growth can give rise to one stable population value, stable oscillations, chaotic behavior, or instability. This is a mathematical result of limits to growth applied to a population capable of exponential increase. This law is not testable; instead, it is a neutral statement of all mathematical possibilities (axiomatic). The law applies at the population scale.

19. Extinction Probability

Extinction of a population is more probable the smaller the population. Three mechanisms operate simultaneously: (1)

smaller populations are more likely to become extinct because of inbreeding depression and concomitant decrease in fitness; (2) random population fluctuations are more likely to lead to extinction when population size is small; and (3) habitat destruction events are more likely to completely wipe out a smaller population. The following equation combines these three causes.

$$P_{survival} = P_{habitat} * P_{inbreed} * P_{fluct} \tag{9}$$

where $P_{survival}$ = the probability of a population surviving without extinction of a local population, $P_{habitat}$ = probability a habitat will not be destroyed, $P_{inbreed}$ = probability that inbreeding will not lead to extinction, and P_{fluct} = probability that random population fluctuations will not lead to extinction. This law states that none of the values of P increase when population size decreases. $P_{inbreed}$ does not apply to asexually reproducing species. This law has been stated previously—the probability of extinction increases nonlinearly with decreasing population size (Brown 1995); however, form of function or breakpoint at which this happens cannot be predicted. Data do suggest very high extinction rates at relatively low population sizes, but common and abundant species can also go extinct (Rosenzweig 1995). This law forms the basis for much of conservation biology with calculation of minimum viable population size being extremely important for long-term preservation of species. This law also holds for establishment of populations. Species introductions of larger numbers of individuals are more likely to lead to establishment of the introduced species (Kolar and Lodge 2001) in part because the larger populations are less likely to go extinct. A potential contingency to this law is that larger populations may be more susceptible to spread of disease. The predictive constraint is related to the probabilistic nature of this law. The law applies to populations at appropriately related spatial and temporal scales of use of environment.

LAWS THAT ARISE FROM EVOLUTION

20. Biotic/Abiotic Interaction

All organisms interact with the environment; they influence it and are influenced by it. Reciprocal influence is a direct consequence of two previous laws, the energy and nutrient cycling requirements. Though the law may seem like a trivial statement, the idea that organisms influence the global environment is relatively recent. For example, a few decades ago, the biotic role of silicate rocks in chemical weathering by controlling deposition of carbonate was established to be involved in global climate regulation (Schwartzman and Volk 1989). The idea that organisms (including humans) can have significant effects on global climate has received substantial scientific study over the last few decades. A predictive constraint is that all organisms influence and are influenced by the environment differently and not all aspects of the environment influence organisms and vice versa. In general terms, this law means that both abiotic and biotic factors must be taken into account by ecologists. This law applies from microbial to global scales.

21. Evolution Affects Ecology

Evolution molds how all organisms interact with the environment and each other (stated differently, both biotic and abiotic processes can drive evolution). History matters to ecology. This law is axiomatic; it is a simple statement of processes that can select among genotypes and phenotypes. This law applies from microbial to global scales.

22. *Specialization*

Organisms are specialized; a jack-of-all-trades is master of none. Physical and chemical constraints dictate that an organism that does multiple things is less efficient than one that specializes (Rosenzweig 1995). No two species will be identical because the process of evolution has a stochastic component; thus, no two species will have identical requirements and competitive abilities (see the first law in the next section). This competitive inequality was formalized by Hardin's (1960) competitive exclusion principle. Plants do not fly; cheetahs have neither time nor surface area to sit and photosynthesize. Prediction by this law is constrained by fairly low levels of prediction because some generalists (e.g., *Homo sapiens*) are very successful. At the coarsest level (primary producer versus consumer, large predator versus small bodied predator, filter feeder versus piercer), the law holds. We live in a world where bacteria, archaea, and fungi are superefficient specialists with regard to unique metabolic pathways and rapid growth; multicellular organisms have unique morphology and specialization of organ systems; animals specialize in predation and behavior; and multicellular plants specialize in obtaining light, nutrients, and defense against grazers. Experiments and phylogenetic tests confirm evolution based on tradeoffs (Kneitel and Chase 2004). Measurement of the degree of specialization of related organisms has been formalized as the idea of niche overlap, and the law has not been disproved by observation of complete niche overlap. This law applies from individual to species scale.

23. Irreversibility of Extinction

Species extinction is irreversible. The mechanism is that rate of evolution is slow, and underlying material for natural selection to operate on is stochastic (random nature of mutation). The probability is almost zero that evolution will produce an equivalent species from a closely related one when a species is lost. It is not completely impossible that replacement will occur, but the probability is exceedingly small. For example, mutants of a simple base pair substitution can revert to wild type, but only rarely do. It is likely that more than one gene is involved in separation of closely related species, making probability of re-evolution of species very low. This law implies that at local scales, local extinction can only be reversed by immigration because the probability of equivalent species evolving is vanishingly small. The law applies at the individual species scale.

VARIABILITY AND ORGANISMS

24. All Organisms Are Unique

No two organisms (species or individuals) are exactly alike because of genetics and environmental heterogeneity. There are genetic mechanisms behind this law. In addition, the physical reality is that two organisms' environments cannot be identical, so the organisms will differ in factors influencing cellular composition or the way stimuli influence behavior. Species that reproduce sexually will always vary; even identical twins vary. The organisms most likely to be identical, clonally reproducing bacteria are likely to have slightly different genomes. In bacteria, organisms with the simplest genomes, mutation rates are around 10^{-4} mutations per gene per generation in the natural environment (Miller 1993). Given that there are 500 to 10,000 genes per bacterium (Casjens 1998), it is likely that there is at least one mutation in one gene per generation of bacterium. In addition, horizontal gene transfer can occur among bacteria, and plasmid numbers are not always exactly the same after division, further decreasing the likelihood of two identical organisms (Miller 1993). Even if two bacteria have identical genomes, they will be at different points of the cell cycle so will have different DNA, RNA, protein contents, and requirements. Multicellular organisms are even more likely to differ than bacteria. Mayr (2004) stated that the property of variation in organisms is what sets biology apart from other sciences. For example, the population view is a constraint that forces biologists to think about mean and variance of populations. In contrast, chemists treat all molecules with the same chemical composition and structure as identical, and physicists treat all electrons as the same. Mayr (2004) proposes that a large part of the scientific revolution created by Darwin is attributed to biologists using the concept of populations. The law that all organisms are unique is part of what sets biology apart from other natural sciences. This law of uniqueness is a constraint to ecological prediction that often can be ignored. For example, many models in population biology assume equivalence of individuals and offer a strong degree of prediction. This law applies at scales of individuals, populations, and species.

25. Population, Resource, and Habitat Heterogeneity

Populations and resources on which they depend are distributed heterogeneously over space and time. Abiotic mechanisms for this heterogeneity are that geological processes and climatic and hydrologic patterns lead to abiotic variation. This abiotic variation and variable species interactions (species interactions must vary if the law that no two organisms are the same holds) among individuals leads to spatial and temporal heterogeneity. This simple observation and the observation that organisms propagate across suitable patchy habitats gives rise to the entire field of metapopulation dynamics. The initial model that provided much impetus to the field was published by Levins (1969) to describe rate of change in occupied patches.

$$Dp/dt = mp(1 - p) - xp \tag{10}$$

where p is proportion of filled patches, m is rate of movement between patches, and x is extinction rate of patches. This model is axiomatic.

One fundamental idea of metapopulation dynamics, assuming a stochastic steady state, is that the probability any patch i will be occupied (J_i) is computed as

$$J_i = C_i/(C_i + Ex_i) \tag{11}$$

where C_i is the probability of recolonization of each patch i, and Ex_i is the probability of extinction within each patch (Hanski 1999). Another outcome of spatial heterogeneity is the concept of corridors connecting larger areas of suitable habitat (Allen and Hoekstra 1992). The predictive constraint is related to the probabilistic nature of this law and the basic spatial or temporal form the variance takes. The law and associated models apply at the population scale.

26. Scaling

Organisms operate across widely varying temporal and spatial scales. Mechanisms for this law are specialization of organisms and genetic drift. Empirically, organisms have an approximate size range of 1 μm to 100 m, and generation times range from hours to centuries. Possibly, lower limits to size of active cells depend on housing the biomolecular machinery that makes cells work. Reasons for upper limits are less clear. Likely, cells grow and divide at the maximal rate given complexity of chemical interactions in cells and limitations of diffusion and transport. As discussed previously, the law that all organisms eventually die is axiomatic. Power functions commonly explain biological patterns across scales (Vandermeer and Perfecto 2006), although the generality of power functions has been contested in specific cases (Alonso and Pascual 2006). Power functions applied to organisms are probably a natural outcome of the diversification of organisms by evolution that leads to size and generation time varying over many orders of magnitude. The fact that organisms operate across such broad scales, coupled with the constraint that physical and chemical laws hold, has led to numerous allometric "laws," or empirical observations, such as metabolic rate scaling to organism size (e.g., Brown et al. 2002, 2004) or the biological consequences of relationships between surface area and volume, size and strength, or size and complexity (Bonner 2006). Constraints on scaling include maximum size of organisms with an exoskeleton, maximum size of mobile animals, and maximum size of flying animals. How to deal with widely variable spatial and temporal scales in ecological systems has concerned ecologists for decades (e.g., Allen and Starr 1982, Levin 1992, Milne 1998). By definition, this law applies across all scales of biological organization and temporal and physical scales under which organisms operate.

BIOTIC AND ABIOTIC INTERACTIONS OF ORGANISMS

27. Species Interact

More than one species is present in all ecosystems, and all species interact with some other species (Lawton 1999). The mechanism of this law is the requirement that species specialize over evolutionary time to be successful. Empirical evidence supporting this law is that no organisms can grow indefinitely with only an input of energy in an otherwise completely isolated environment. No cases have been documented in which a single species or strain lives in isolation (even species grown in isolation by humans are interacting with another, the person who maintains the culture). As far as is known, this is a strict condition and a consequence of the requirement that species specialize. The most obvious general case is that heterotrophic organisms need autotrophs to provide food and autotrophs depend on heterotrophs to recycle nutrients. The predictive constraint is inability to predict the proportion of species with which each species in an ecosystem interacts. This law forms the most fundamental principal of Odum's (1959) popular and influential ecology text and applies at the species scale.

28. All Types of Reciprocal Interactions Are Possible

If any two organisms are drawn randomly from a community, it is possible that the direct interactions between them take any sign (+, −, or 0). The mechanism is the specialization of species because there is no other reason to assume all interaction types must occur. This statement provides the basis for community interaction structure, and the question then becomes (Dodds 1997), what is the relative distribution of these 0, +, or − interactions, and how are they coupled? Reciprocal interactions then can take the form 0/0 (neutralism), +/+ (mutualism), −/− (competition), 0/+ (commensalisms), 0/− amensalism, or +/− (exploitation). A brief note on terminology is necessary here. Mutualism is sometimes termed *symbiosis*, but I support the view that the latter term means organisms living close together regardless of interaction type. All mutualists are not symbiotic (e.g., plants and their pollinators), and all symbionts are not mutualists (e.g., parasites). Also, the term *exploitation* is more general than the commonly used predation. Parasitism, pathogens, herbivory, and other exploitations (e.g., an epiphyte growing on a tree that intercepts light and nutrients so affects the tree negatively but receives a place to live) are often not considered predator/prey relationships, but the relationship is covered with the term exploitation. A constraint in this law is that interactions are variable, so there may rarely be a true zero interaction, and interactions may oscillate between + and − over time. This law applies at the species scale.

29. Diversity of Interspecific Interaction

All communities (all organisms found in a specific area) will have +, −, and 0 direct interactive effects among species. Interaction diversity is a result of the laws that organisms specialize and diversify and the law that organisms interact. If the statement that "no organism is capable of complete nutrient cycles" holds, all organisms directly or indirectly need each other. Thus, there must be at least some positive interactions present. If autotrophs are present and heterotrophs consume them, at least one negative direct interaction is present. Given the widely divergent spatial and temporal scales used by organisms and diversity of all known areas on Earth, there is an extremely high probability that some organisms will not interact with each other directly. Because there is generally more than one organism trying to use similar resources, competition is also likely (no community in which competition does not occur has been described yet). This law applies at the community scale.

30. Variance of Interspecific Interaction

Not all organisms have the same strength of interaction, so some organisms will be relatively strong interactors. This law is a consequence of specialization of species and the law that no two organisms are exactly alike. However, what is not known is how interaction strengths are distributed in natural communities (e.g., does the distribution center on 0, + or − interactions, and are distributions normal, skewed, or discontinuous?). We can predict there should be a distribution of interaction strengths from basic principles, but prediction is constrained by not knowing the shape of distributions. The idea of keystone species (Power et al. 1996) arises from the fact that it is likely a few species are disproportionately strong interactors relative to their abundance. The law applies at the community scale.

31. Competitive Exclusion

Given that two species are limited by the same resource and only one of them can be the superior competitor (variance of interactions, all organisms are unique), only one of them will be able to coexist under the constraint of equilibrium conditions and lack of limitation by other factors such as predation or mutualism (Hardin 1960). There are several constraints on this law. First, the system being considered must be at or near equilibrium. Second, resources must limit population growth of species (not predation or some other factor). Third, if interspecific competition is stronger than intraspecific competition, only one species can dominate. If intraspecific competition is stronger than interspecific competition, coexistence is possible. This outcome is a result of the mathematics of Lotka-Voltera competition models but does not necessarily depend on the specific mathematical form of density dependence given by the logistic equation. The law can be demonstrated under closed conditions (e.g., cultures of algae), and has not been disproved under such conditions (well-mixed environment and a single limiting nutrient). The law applies at the species scale.

32. Linkage of Interactions

Indirect interactions (interactions between two organisms mediated by one or more other organisms) allow propagation of interactions through communities. This law arises from simple mathematical properties (i.e., it is axiomatic). The axiom is, if A affects B and B affects C, then A will have an effect on C mediated by B, even if loops are not allowed to involve a species more than once (Fig. 1, assuming allowable interaction chains with no loops). This law leads to a terrific number of potential interactions as number of species increases. If I_c is total number of interactions in a community, the potential number of indirect interactions with n links is

$$I_c = S*(S-1)*(S-2)*\ldots*(S-n) \qquad (12)$$

In this case, the total possible link length, n, needs to be one fewer than the number of species (S). Given a group of S species, there

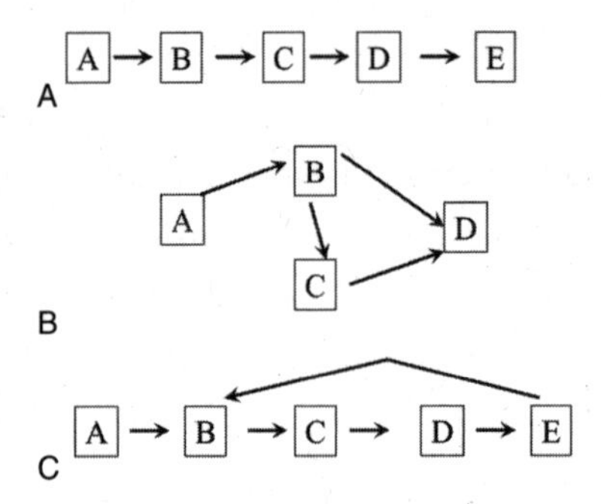

Figure 1. Examples of indirect interactions. If it is assumed that each species can participate in a chain only once, (A) represents one interaction chain between A and E, (B) shows two interaction paths from a to D, and in (C) the effect of A on B is only direct because it cannot pass through B twice (e.g., A to E and back to B).

are $S^*(S - 1)$ possible direct interactions, or interactions with one link. However, the number of possible indirect interaction chains with n links increases dramatically as n increases. Equation 12 gives rise to a huge number of potential interactions (Fig. 2). The most obvious manifestation of this law is the existence of food web dynamics, but interaction webs do not need to be based solely on predator-prey and grazer-autotroph interactions. A predictive constraint is an a priori characterization of how interactions propagate through communities. Indirect interaction strength seemingly should decrease with the length of the interaction chain and not be stronger than the weakest interaction, but this is not necessarily the case (Berlow 1999).

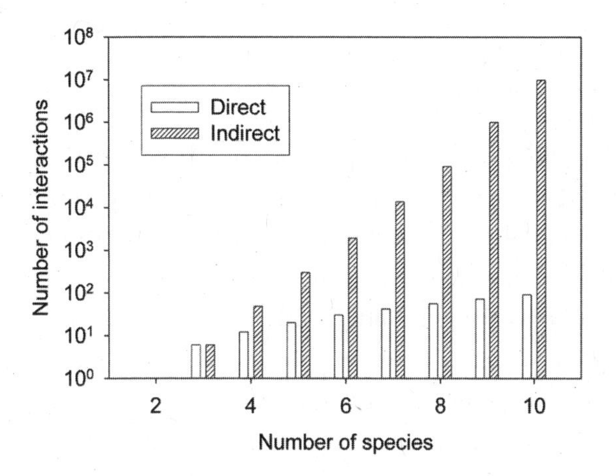

Figure 2. Total potential number of direct and indirect interactions as a function of community size.

33. Nonpropagation of Interaction Chains

A zero interaction stops an interaction chain. Propagation is a matter of definition and is axiomatic. A consequence of interrupted chains by zero interactions is that divergent spatial and temporal scales of operation make some zero interactions likely, so the fact that a zero interaction stops an interaction chain suggests it is extremely unlikely that all conceivable interaction chains among organisms in a community are present. A predictive constraint is that the probability that an interaction is exactly zero is not known. This law can be stated differently; communities have patterns of interactions and are not homogeneous. This law applies at the community scale.

34. Heterogeneity Increases Diversity

Spatial variation within an area leads to an increase in diversity relative to a spatially homogenous area. The mechanisms for this greater diversity are the laws of spatial variance in abiotic factors and evolutionary diversification. Temporal variation (e.g., changing disturbance frequency) does not necessarily lead to a similar pattern. Actual boundaries of spatially heterogeneous habitats have specific properties different than properties found within any of the areas that abut the boundary (Strayer et al. 2003). Therefore, increased diversity may be due to both specialization to different individual habitat types and specialization to the interface between habitats. Also, heterogeneity may alter refugia for prey. Some constraints are that the law is scale dependent (i.e., heterogeneity of habitat needs to be relevant to the species being considered) and similar areas need to be compared (e.g., an equal area of homogenous profundal marine sediment can have much greater diversity than an equal area of heterogeneous desert). This law applies from small spatial scales (e.g., sampling microbes over areas of square centimeters) up to global scales.

35. Diversity Positively Correlated With Area

Larger areas will have more species. A contingency is that similar types of environments must be compared for the law to hold. A large area of an ice cap and a small area of tropical forest cannot be compared for this law to hold. Mechanisms are the law that spatial variation leads to an increase in predicted diversity and probability of a greater range of habitats increases with increase in area. This pattern was noted in the late 1800s (Rosenzweig 2003) but has been confirmed with multiple taxa over numerous areas (Rosenzweig 1995). The pattern even extends to bacteria

(Horner-Devine et al. 2004), so it seems likely to apply to all organisms. The relationship is based on

$$dS/dt = O_s + I_s - Ex_s \qquad (13)$$

where dS/dt = change in species numbers over time, O_s = speciation rate, I_s = immigration rate, and Ex_s = extinction rate. An equilibrium point, at which dS/dt is zero, will lead to a greater number of species as area increases through several mechanisms. First, evolution rate and immigration rate will be greater with more spatial diversity, and total spatial variation is greater with greater area. Second, extinction probability is greater with small populations, and smaller areas can only support smaller populations. The shape of the often-observed curve (saturating) is not predicted by this law. The law that covers saturation is that there are fewer species per unit area in larger areas. This is a more difficult law to provide mechanisms for, so it will be approached in the section on species area curves in Chapter 2 but not proposed as a fundamental law. Other mechanisms for the power relationship behind species-area curves, such as "species-level fractals," are controversial (Ostling and Harte 2003). This law applies from small spatial scales (e.g., sampling microbes over areas of square centimeters) up to global scales.

SOME CANDIDATE LAWS

1. All Organisms Have Diseases (Viral or Otherwise)

Apparently viruses have been present since early in the evolution of organisms. Given the selfish nature of genes, rogue information sequences (either as escapees from existing genomes or as drastically reduced parasites) are probably inevitable. The requirement for parasites (but not predators) has been proposed previously (Wilkinson 2003).

2. Organisms Will Evolve the Ability to Tolerate, and Likely Metabolize, Unique Organic Compounds

The mechanism is taken from evolution and species interacting. To survive, organisms must be able to withstand chemical challenges, particularly in the microbial world. This potential law is of practical importance because we can predict that for every organic pollutant, we should be able to find a bioremediation tool; for every antibiotic, a resistant organism will arise; and for every pesticide, resistant pests will surface. Similarly, we can also predict a coevolutionary arms race between herbivores and plants. The contingency is that some highly recalcitrant compounds can be broken down slowly or not at all (e.g., plastics, organic nanomaterials). Some organic toxins may always incur a cost to most organisms. Although the process of bacteria adapting to break down organic pollutants is predictable in a general sense, how soon it will occur and the specific pathway of breakdown is difficult to predict (van der Meer 2006).

3. Nonequilibrium Nature of Life Leads to Nonequilibrium Conditions in Ecological Systems

This potential law is based on the idea that there is a mixture of order and disorder in biological systems. For example, cells are fundamentally perched between complete order (i.e., stasis) and disorder. A similar law has been proposed as a candidate law for development of our biosphere (Kauffman 2000). An argument has been made on thermodynamic grounds that at 0 K there is no order or disorder, and as temperature increases, processes that create order can occur more quickly, but the cost of maintaining structure increases. Thermodynamic arguments can be taken to be the reason that life can exist only within a narrow temperature range (Jørgensen and Svirezhev 2004), although this range also coincides with the range in which water is liquid under standard atmospheric pressure. Thus, lipids must remain fluid to operate, but the lipid bilayer must retain integrity for cells to maintain their cell membrane. Consequently, individual organisms are, by requirement, nonequilibrium systems, so populations, communities, and ecosystems containing individual organisms are, by extension, nonequilibrium systems. Solé et al. (1999) note that complexity in evolution and ecology is a natural product of systems that are close to instability. Allen and Hoekstra (1992) make the case that ecologically interesting phenomena occur in areas where prediction is difficult (where there are switches in fractal dimension). Organisms, by default, operate across fractal dimensions and may be bound by nonequilibrium conditions for this reason.

4. The Law of Failure

Omerod (2005) claims that it is an "ironclad law" that species, political policies, and economic entities go extinct. This potential law is closely related to the law that extinction is irreversible. With respect to biology, he claims that the law is 99.99% true, which begs the question, is 99.99% "ironclad?" The vast majority of all species have gone extinct. However, cyanobacteria are still present after 3 billion years. The failure law can be considered the inverse of natural selection. If you can predict which species will succeed, you can also predict which species will fail. The random nature of selection makes such prediction imprecise at best. I previously proposed the axiomatic law that all individuals will fail (die). The death law applies to individual organisms but not necessarily to all their genetic sequences. Heisenberg's uncertainty principle may dictate that all information systems based on quantum mechanical principles will ultimately fail (e.g., the molecule DNA will mutate randomly because information transfer cannot be perfect). Extinction or extirpation of specific organisms over ecological time within a specific environment is only predictable if substantial information is available, and extinction is rare over time periods experienced by humans, so the degree of contingency of this law is considerable. The fundamental principle of human influence does lead to the prediction that human activities will tend to produce extirpation and extinction, a prediction borne out by current reality.

5. Consumer-resource Oscillations as a Law of Populations

Or, "a pure resource-consumer system will inevitably exhibit unstable oscillations" (Turchin 2001, p. 21). This candidate law is a restricted subset of the law *Population stability not determinant* as proposed earlier. Although this is a mathematically sound argument, the candidate law offers modest predictive ability in real ecological situations. Oscillations are only observed mathematically when specific conditions are met. While these conditions can be met in natural systems, it is not always clear when they will arise, and reasons for the oscillations in some real world cases remain obscure. Take for example the classic observation of population cycling in the snowshoe hare. At times, predation pressure can explain cycles; at other times, availability of food can explain them. Even after over 25 years of research, some mechanisms (e.g., a mechanism for the observation that low populations of hares can remain for several years in spite of ample food) are not understood (Krebs et al. 1993). Admittedly, this is an example of how much contingency is acceptable for a pattern to be deemed a law, and it is my subjective view that the degree of contingency is too great for this law to hold. Peter Turchin (personal communication) disagrees.

SOME USEFUL GENERALIZATIONS OR PATTERNS

Here are some potential laws that I do not think should be considered laws, but could qualify as useful generalizations or patterns.

1. Leibig's Law of the Minimum

This is considered by some an ecological law. If a system is homogeneous and a process requires multiple resources in a fixed proportion over time, the resource that is available at the lowest relative rate will limit the process. Even individual cells in a well-mixed chemostat-continuous culture, however, have varying requirements because they are at different points in the cell division cycle (e.g., DNA synthesis has a large P demand). There are many cases in which a nutrient is supplied at a low-rate relative to others so all primary producers are limited by a single nutrient, but predicting when this will occur is difficult. Even in continuous culture of a single species of microorganisms, more than one nutrient may limit growth (Egli 1991, Egli and Zinn 2003). Odum (1959) felt that the most important aspect determining the applicability of Leibig's law is the existence of steady state conditions. If the laws proposed previously are accepted, the law of the minimum does not hold. There are numerous cases where assemblages of organisms are balanced between limitation of more than one nutrient in equilibrium (Tilman 1982) and nonequilibrium systems (e.g., Dodds et al. 1989). Organisms should evolve the ability not to be constrained by limiting resources, leading to limited predictive value of the law of the minimum. Allen and Hoekstra (1992) suggested using the concept of constraints rather than limiting factors. Using "multiple constraints" could avoid invoking the assumptions of Leibig's law that could be assumed using the term *limitation*.

2. Occam's Razor

Also know as the principle of parsimony, this is commonly invoked in science. However, the simplest explanation is not necessarily the correct explanation. The inapplicability of the simplest explanation is particularly common in ecology, where evolutionary history dictates constraints on how organisms solve ecological problems. Convergent evolution of different structures to solve similar problems is an example of different solutions that are not always the most parsimonious. Organisms fly or glide using wings and feathers (birds), skin attached to forelimbs (bats), skin stretched between forelimbs and hind limbs (flying squirrels), or flaps of skin along extendable ribs (flying dragons, *Draco*, species of gliding lizards and gliding snakes). None of these structures originally evolved for flight, and some are simpler than others. The most complex (feathers) are also the most efficient solution.

3. Shelford's Law of Tolerance

This law states that too little or too much of any factor can limit an organism (Odum 1959). Too much light, too high a temperature, too extreme a pH, or many other extreme factors will limit individual organisms. Some factors, such as toxins, cannot be present in too small concentrations. With respect to gradients of factors where optima do occur, there is not one simple explanation for excesses leading to inability to survive. Maybe my argument for rejecting this law is in violation of my argument that multiple mechanisms can be used as the basis of a law. But, the mechanisms are so varied that the degree of contingency seems too great. Temperature and light extremes lead to molecular degradation, extremes of water may cause terrestrial organisms to lose the ability to take up oxygen (drown) or to become too dry to survive, and extremes in toxic chemicals may interfere with specific molecular processes. Extremes of pH or redox can interfere with homeostasis or lead to denaturizing proteins. This pattern does not allow for generality, so I do not consider it a law.

4. If an Organism Regularly Displays a Characteristic That Costs It Energy, It Must Have Some Adaptive Value

Activities that cost energy may have adaptive value, but they may also be imperfect solutions to problems of survival or may have had adaptive value previously. Therefore, aspects of organisms that cost energy are likely to have adaptive value, but this is not a law, just a strong prediction. Assuming this to be a law is why injudicious application of optimal foraging theory can run into problems.

5. The Allee Effect of Intraspecific Cooperation

From Berryman 2003. Although many good examples of how intraspecific cooperation as well as fundamental reasons cooperation will benefit individuals are available, there is no way to use this to predict any specific cooperative behavior or if it will occur for every species. Many organisms, such as unicellular microbes (not including the slime molds), should not exhibit this effect. Therefore, the Allee effect is a pattern to look for in population ecology, but it is not a law.

6. Self-limitation of Population Size Is a Fundamental Law of Populations

From Turchin 2001. Because so many other things can limit population size, we cannot predict when intraspecific competition will limit population growth in real systems. Under constrained conditions, this holds as a law (for example a unialgal culture growing in a flask). Something must ultimately limit population size, but not necessarily intraspecific competition or regulation. I prefer the more general statement of the related law, *limits to growth,* advanced previously; no population can increase indefinitely.

7. Biomass Pyramids Will Occur

Occurrence will be in the form of a broader base (greater biomass) at lower trophic levels. This generalization is based on principles of thermodynamics, but it assumes that productivity is always proportional to biomass. There are several reasons this idea does not hold across a number of ecosystems. One problem is that it is difficult to define trophic levels because they are a fuzzy concept with omnivory. A second, more serious criticism is that production is not always completely tied to biomass. In terrestrial systems, structural material is required for successful competition for light, so the biomass of primary producers is generally great. However, in aquatic systems, primary producers or basal consumers can have minimal amounts of structural materials and turnover rapidly. Rapid turnover low on the food web can lead to inverted biomass pyramids. Such an observation has led to the Allen paradox in streams, where it has been demonstrated that biomass of secondary and higher consumers exceeds that of primary consumers. The paradox can be explained on the basis of high production per unit biomass of primary consumers (Dodds 2002).

8. The Law of Exergy

Exergy is the opposite of entropy and is the law that ecosystems tend toward maximum exergy storage, maximum power, and maximum ascendancy as an emergent property (Jørgensen and Svirezhev 2004). This law is problematic because it is difficult to ascribe any evolutionary tendency to individual ecosystems. It is true that primary succession leads to more complex communities that hold more information, but not all ecosystems move this way. The ideas expressed in this law are related to the concept that life is a self-organizing system that forms, regulates, and ensures the dynamic stability of the environment (Kondratyev et al. 2004). Related candidates for laws are discussed by Kauffman (2000) who concludes the evidence is not strong enough yet for them to be accepted as laws.

9. Organisms on Earth Have Evolved Toward Regulation of Global Climate

Also know as Gaia, a result of group selection. This law in its strong form, is probably not accepted by most ecological scientists, rather it is more of an ecological philosophy related to the concept of "deep ecology." The strong form of this law is that a global ecosystem that does not self-regulate will go extinct in its entirety, so the global level of selection leads to planetary regulation. The weak form can be covered by the candidate law "*Organisms will evolve the ability to tolerate, and likely metabolize, unique organic compounds*" presented in the section preceding this one. Though self-regulation of global climate is now widely acknowledged, Volk (1998, p. 239) sums it up best. "What organisms do to help themselves survive may affect the planet in enormous ways that are not at all the reasons those survival skills were favored by evolution." I will discuss a weak form of self-regulation in Chapter 2 when considering the theory of complete nutrient cycles.

THE PROMISE OF REDUCTIONISM IN SOLVING ECOLOGICAL PROBLEMS

There is a danger in assuming that fundamental physical and chemical laws and mathematical axioms will be able to describe all the important features of ecological systems (i.e., that there is nothing unique about ecological systems). A colleague of mine who specialized in virology once suggested the only course requirement for all graduate students in biology should be biochemistry, all else follows from that. My counter to his proposal was that all molecular biology students should be required to take quantum mechanics because it describes the most fundamental properties of matter and thus explains biochemistry. The problem is that the fundamental physics of atoms cannot (at least not yet) be used to describe the behavior of complex biological molecules. Well-known physicist, Richard Feynman (1999) noted that all the equations for atomic and molecular forces in water cannot accurately describe turbulence. He stressed how little physics really can explain about the world. Likewise, it seems unlikely that biochemistry and molecular biology can completely explain ecology.

The problems with reductionism are well explored by Stuart Kauffman (2008). He makes a clear case that emergent properties exist that cannot be explained by reductionist approaches. He singles out biology because reductionist laws cannot be used to predict the course of evolution. This argument echoes that of Mayr (2004). In Kauffman's (2008) *Reinventing the Sacred: A New View of Science, Reason and Religion*, the potential for emergent laws is explored, but none of Kauffman's candidates for laws are any more than hypotheses at this point. The exploration of emergent behaviors is a relatively new field and may ultimately yield substantial predictive value.

A more realistic alternative approach to that of extreme reductionism seems to apply the principle that the mechanisms to predict what occurs at one level in the biological hierarchy (e.g., organism, species, population, community, ecosystem) tend to lie at the level of organization below that. In Kauffman's (2008) view, this means that the arrows of explanation all point down the hierarchy. It is fairly clear that all the arrows do not point down. Although there are numerous counter examples (e.g., many ecosystem properties are based on fundamental chemical properties such as the solubility of P and iron compounds or the inertness of the chemical bond in gaseous N), it is generally difficult to extract explanations and predictions from causes far removed from the phenomenon of interest.

The observation that mechanisms tend to be related to the next lower level of organization does not merit the status of a law because there are many exceptions. These exceptions range from the global ecosystem influencing local populations to chemical effects (e.g., acid precipitation) influencing communities. Ecologists have wrestled with this problem when considering the ability of small-scale experiments to explain whole-system behavior. A good case has been made for bottle experiments providing poor predictive power at whole-lake ecosystem scales (Schindler 1998), indicating that reductionism is, at best, difficult to apply correctly to ecological studies.

The strongest arguments against reductionism come from the observation that emergent properties may be important features of ecological systems. Emergent properties are those that arise only as a feature of the systems as a whole and cannot be predicted by the properties of the constituent components. Such emergent properties include the observation that order arises when energy is forced through systems (dissipative systems, Schneider and Kay 1994), and the inspection of the way that a pile of sand grains builds as more sand is added to the top of the pile (Milne 1998). Possibly, many emergent behaviors of ecological systems occur because ecological systems are near critical points, at which details of specific interactions matter less than the paths that interactions propagate across (Stanley et al. 1996). However, we are a long way from using emergent properties to make general predictions about behavior of ecological systems other than the point that such properties may exist.

When considering patterns that occur in biological systems, it seems probable that explanations are most likely to arise at the level below that of interest, with decreasing probability that this will occur as you move further away in level of organization. However, levels of observation may simply be constructs of the observer and not real. Furthermore, links across levels may occur commonly (Allen et al. 1984). Although all this means it is unlikely, or at least difficult, to make predictive laws that operate across levels of a hierarchy, matters of scale are clearly important.

The approach taken here is not reductionist in the sense of always looking for explanations at the lowest level possible (e.g., genes are taken as the ultimate explanation, as is thought by many biologists today). Rather, it is reductionist in attempting to apply the simple set of basic laws proposed, combining the laws with strong patterns (e.g., microbes dominate detritivorous activities in ecosystems) and seeing how far they can be taken to assemble theories in ecology. Chapter 2 illustrates how the basic set of proposed laws can be used to construct ecological theories.

CHAPTER 2

Theories

A theory is a plausible or scientifically acceptable general principle or body of principles offered to explain phenomena. The topic of theories in ecology, their relationship to modern philosophy of science, and their utility are explored elegantly by Pickett et al. (1994), and only a few issues covered in that treatment are discussed here. In general, Pickett et al. (2007, p. 63) define a theory as "a system of conceptual constructs that organizes and explains the observable phenomena in a stated domain of interest." The theory then allows for causal explanation of phenomena in the domain of interest. Theory, as they approach it and as discussed in this chapter, does not mean hypothesis, as it often does in colloquial use.

Theories can explain phenomena with tighter or looser constraints. On one hand, a theory can lack predictive power if it does not take enough constraints or contingencies into consideration. On the other hand, if a theory takes too many parameters into account, it can "overfit" the data. A complex model may fit a limited data set well but not be useful in a general sense because constraints narrow the range of conditions that can be predicted (Ginzburg and Jensen 2004). Similar trade-offs were discussed in the Introduction with respect to laws.

Uncritical scientific approaches can lead to unwarranted acceptance of theories once entrenched. Buller (2005) gives the example of a chemistry student in a college laboratory. If an experiment does not work, the student does not assume that the theory that is the object of the laboratory should be discarded; they simply assume the experiment was done incorrectly. If the experiment provides close to the correct results, any error is assumed to be the error of the student, but the experiment is still assumed to confirm the theory. Such an approach could elevate a theory to general acceptance before it is deserving of such status.

The ecological theories explored in this section are ideas that have been explored by others, but hopefully foundations of these theories are made more explicit by using the laws proposed in Chapter 1. Concepts termed probabilistic laws by Pickett et al. (1994) are generally considered as strong patterns in my framework. This chapter does not present all potential theories in ecology, rather it focuses on some that make particularly good examples or those that are broad enough to overarch a variety of ecological systems.

These theories are not all of equal importance (i.e., some have a much broader "domain"). A theory of nutrient cycling in streams is less general than the theory of evolution. My aim is to provide a group of examples of theories, and areas I am more familiar with allow me to provide better examples. Theories of evolution, nutrient cycling, and community interaction are at the broadest levels. Others are more specialized. Hopefully this approach will provide a general avenue for others to consider how to construct theories that are more specific to subdisciplines in ecology or at least how to think about assumptions that underlie their subdiscipline.

A problem with ecological theories is definition of terms for ecological units. There is no universally agreed on definition of terms such as *population*, *community*, or *ecosystem* in ecology (Jax 2006). These terms can be defined with general concepts or statistical approaches. For example, a population can be defined statistically by gene flow rates or simply as all the individuals of a species in a particular area. Arguments over semantics are mostly avoided, and in the book, I use terms in the operational (conceptual) sense in a fashion generally consistent with common usage by ecologists.

Theories have a structure and are built on other theories. Here, the approach delineated by Pickett et al. (1994) of open systems of understanding is taken. That is, theories are based on nested structures that support the basic theory (Fig. 3). I do not propose a general theory of ecology because all the constituent theories are not well defined, although a compelling argument for a unified theory of ecology has been made (Scheiner and Willig 2007). My first proposed law (an umbrella approach that all chemical, physical, and mathematical laws hold) is used as the structural framework for the theories here without being explicit about their contingency or their inapplicable aspects. For example, a negative species population size is possible mathematically but not logically. Still, we do not discard all mathematics. Although the text is locked into a linear presentation, there are more complex relationships among theories. The theory of ecosystems, for example, can be thought of as having a variety of nested theories. The subtheories are subject to all constraints of the overarching theory, but by adding additional constraints, improve predictive ability. For

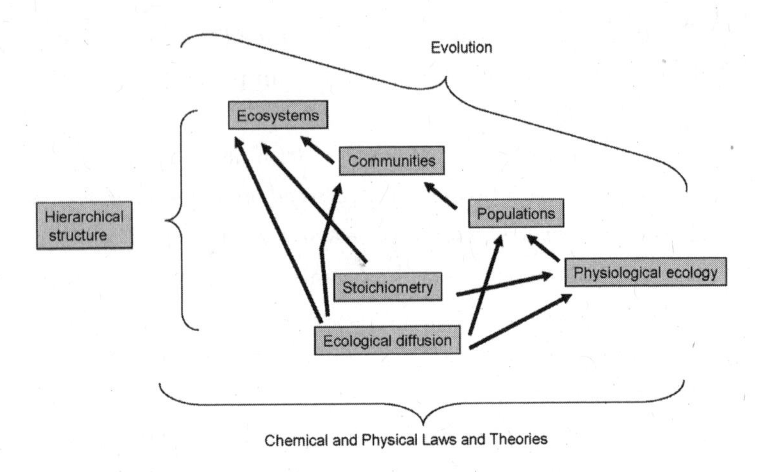

Figure 3. One potential arrangement of disciplines ecology as discussed in this book. There are other potential levels (biomes, biosphere) that are not explicitly considered here. Evolution reacts reciprocally with all levels, and physical and chemical laws constrain all levels. Some cross-cutting theories, such as hierarchy theory, the theory of stoichiometry, and ecological diffusion, influence multiple levels.

example, the theory of island biogeography adds to community theory the constraint that there are insular habitats that vary in rate of colonization from outside.

First, each theory will be introduced with pertinent brief introductory material. Second, the laws proposed in Chapter 1 that specifically apply to the theory being discussed are listed. Most of these theories also depend on strong observations that do not quite qualify as laws but allow for better predictive ability; these are listed third. These strong patterns are often empirical observations with mechanisms not well enough established or too limited by constraints to be considered laws.

Fourth, I discuss some major predictions based on the theory and limits to prediction.

CROSS-CUTTING THEORIES

Major subdisciplines of ecology can be arranged hierarchically into levels of organization (Fig. 3). Each level of organization influences the one above it. Other theories, such as ecological stoichiometry and ecological diffusion, influence multiple levels on the hierarchy. Those cross-cutting theories or concepts are discussed first. In the following sections, theories influencing each of the subdisciplines are discussed starting with physiological ecology, then moving to populations, communities, and ecosystems.

Hierarchical Structure

One of the most fundamental concepts that students learn in introductory biology courses is a hierarchical view of biological organization. The classification of molecular, cellular, multicellular, organs, organisms, populations, communities, ecosystems, and biosphere can be criticized based on overlap of the various levels. However, the theory offers ways to classify and subdivide ecological systems into parts amenable to study and determination of mechanism. Deeper insights related to hierarchy are possible. The issue of a general theory of ecological scaling is approached in Chapter 3.

Laws and Theories Most Applicable System openness, Scaling.

Strong Patterns (This list is presented as fundamental principles by Allen and Hoekstra (1992). I view them as strong patterns rather than laws because of their exception rate).

- Patterns at one scale are most likely to be influenced by patterns at similar scales.
- Higher levels of organization have longer return time (low frequency).
- Because higher levels of organization have longer return time, they provide the context for lower levels. For example, evolution can select for organisms that are able to predict the context at the next higher level of organization.
- In nested systems, the whole system turns over more slowly than its component parts.
- Upper levels constrain levels below them.

Some Predictions Probably the greatest strength of hierarchical structure is that it guides ecological investigators toward where to look for context and components of an entity of interest. For example, in the case of endangered species, a conservation investigator would try to establish if context (habitat) is being maintained. The investigator would also want to be certain that the underlying population biology of the species that maintains the required genetic diversity is intact. If a species is still in decline and habitat appears intact and population structure is adequate, the investigator may start to look elsewhere for factors influencing the decline. The investigator could concede it is futile to study the molecular biology of the species because so little is known about rare species and because they know that molecular characteristics are somewhat removed from species conservation and not amenable to management. Likewise, they would not worry about global effects (although they may be interested in the long-term influence of global climate change), because they are far removed from the system of interest. The investigator would be more likely to look at predation or disease because these factors are more closely related to the hierarchy of interest (population persistence).

Defining the appropriate scale of investigation is difficult, but hierarchy theory provides the first step toward explicit recognition of the importance of scale. Most ecologists unconsciously take this approach to study of their system of interest, probably because of their prior experiences with ecological hierarchy. Specifically defining the scales of interest and how the scales interact can help guide broader understanding (Lowe et al. 2006).

Finding scale-invariant patterns may be useful even if the mechanism behind the patterns is not known. Deviations from known scaling relationships can be used to indicate ecological disturbance. For example, scaling of distributions of tree stem size is altered with fire suppression (Kerkhoff and Enquist 2007). Such analysis of empirical patterns may be particularly useful for predicting the influence of global environmental change on species, because it can deal with large spatial scales that are relevant to global phenomena (Kerr et al. 2007). If the mechanism behind the scaling is not known, the change in pattern may not be easy to interpret.

Once general network properties are known, these patterns can be used predicatively in similar networks (Clauset et al. 2008). In this case, links in poorly resolved networks can be predicted based on other well-resolved networks. This is, however, an empirical approach to prediction. If basic mechanisms are known that underlie network structure, then hierarchical structure could be predicted based on first principles, rather than similarity based on shared properties with other networks.

Spatial and temporal scales are broadly linked across systems allowing investigators to concentrate on causal mechanisms. For example, redwood trees have long generation times, and climate patterns over years or decades might drive growth and reproduction. Daily variation has minimal influence on the biology (except perhaps a large fire or a day of logging). In contrast, the microbial population in the soil below the redwood tree will respond readily to diurnal, daily, and weekly fluctuations. Geological processes operate on timescales appropriate to biogeography and evolutionary pattern. Small-scale turbulence links to coral reef physiology of individual polyps, yet multiple year patterns (e.g., El Niño) also influence growth of entire colonies of reefs.

Some Difficulties with Prediction Sometimes ecological activities that occur at a lower level of organization can constrain features at a higher level of organization. The occurrence of "ecosystem engineers" illustrates activities that move up the organizational hierarchy. For example, beavers (*Castor canadensis*) can create many dams in a high mountain valley that fundamentally alter geomorphology by trapping sediment and building up the flood plain. Many centuries of beaver activity can create a large wetland area where flow is dispersed. This system allows flow to spread out and modifies the effects of flooding and dry periods on downstream organisms. The plant community (wet meadows, willows) is driven by this single species. Furthermore, biogeochemical processes that occur in anoxic sediments, such as methane production, are stimulated.

Evolutionary response to local conditions of light availability and predation pressure to competing plants ultimately influenced global biogeochemistry. Evolution of degradation-resistant compounds that protected trees from herbivores in ecological time allowed trees to develop large structures for vertical extension and competition for light. Growth, death, and deposition of remains of large woody vegetation built of degradation-resistant lignin, led to vast deposits of coal and oil stored for geological time periods (Wilkinson 2006). Ultimately, structures that had originally evolved as a response to local conditions led to alteration of global biogeochemical patterns.

Thus, patch size is context (organism)-dependent, making prediction more difficult. A model of hierarchical patch dynamics was proposed (Kotliar and Wiens 1990) and has been heavily cited, although mostly by review papers rather than researchers who have refuted or supported the model. This model suggests that the lower-level patches are simple (homogenous within the patch) and have well-defined boundaries as perceived by the organisms that use them. At more complex levels higher in the hierarchy, patches have more internal structure and more diffuse boundaries. At the smallest scale (the grain), organisms perceive the patch as heterogeneous. The upper scale of patches that the organism responds to is the extent. At the extent level, the organism perceives much heterogeneity. If this classification of patches is true, it means that patch size is dependent on the perception of the organism (something that is difficult if impossible to know) and thus difficult to predict.

Determining patch size, or more generally, the characteristic length, in ecological systems can be difficult. Problems exist with the methods used to make these determinations (Habeeb et al. 2005). Particularly, systems with complex self-organization (i.e., most ecological systems) can exhibit multiple characteristic lengths. Kotliar and Wiens (1990) suggest as scales increase, patches have more heterogeneous structure and fuzzier boundaries. Indistinct boundaries and heterogeneity are not amenable to prediction.

Ecological Stoichiometry

Ecological stoichiometry is defined by Sterner and Elser (2002) as the balance of multiple chemical substances in ecological interactions and processes, or the study of this balance. They note the fundamental requirement of all organisms for roughly the same proportion of elements and how different organisms can vary these proportions. The primary chemical and physical laws that this theory is based on are the conservation of mass and the law of constant chemical composition (all samples of a chemical compound have the same composition). The theory of ecological stoichiometry is also based on the law that organisms affect their environment and the environment affects them. The theory (in my view) states that organisms have certain broadly similar requirements for elements (Fig. 4). These organisms interact with the environment and each other based on fulfilling these requirements. Organisms have differing abilities to be plastic in their stoichiometry, and variable requirements lead to richness in predicted responses to proportions of elements available in the environment.

Laws and Theories Most Applicable Laws from physics and chemistry. System openness, Recycling rates, Energy requirement, Nutrient cycling requirement, Maximum metabolic rates, All organisms are unique

Strong Patterns

- All species require a similar basic cellular elemental balance, because they use the same molecular and biochemical machinery.
- Primary producers have more plastic stoichiometry than consumers.

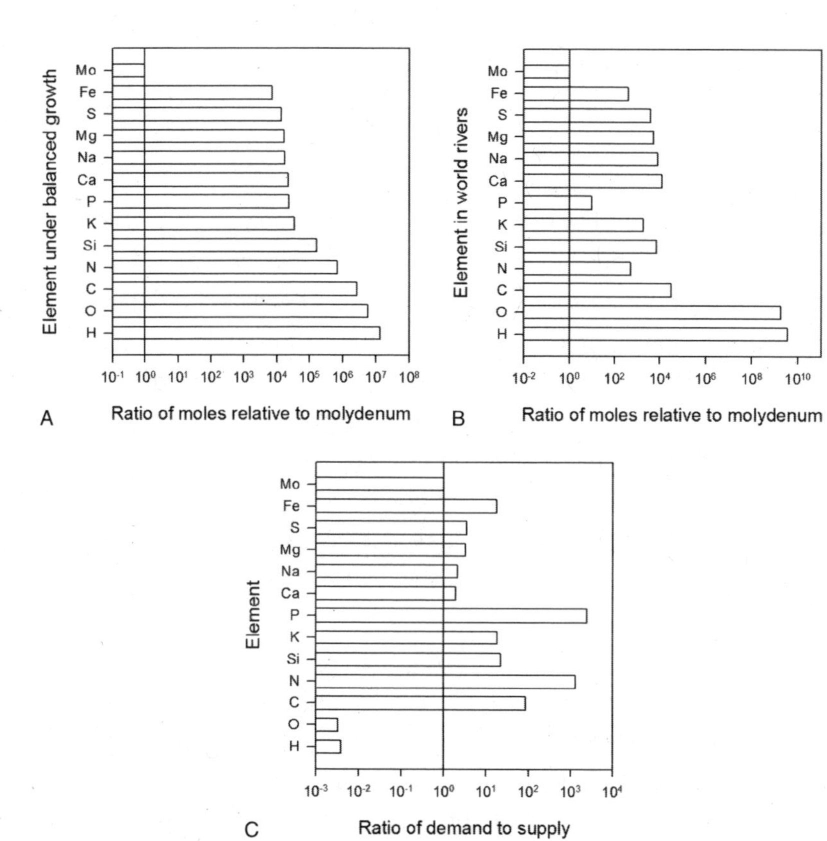

Figure 4. Elemental composition of aquatic algae and plants (A) compared with availability in freshwater (world rivers, B). All composition data are in moles or atoms relative to molybdenum. Average demand/supply (C) is algae and plants divided by rivers. [Data compiled from literature by Dodds (2002)].

- Organisms with more plastic stoichiometry will more strongly reflect the composition of their food sources in the composition of their tissues.

Some Predictions The theory of ecological stoichiometry can make reliable predictions about the physiological characteristics of nutrient processing by organisms. If the range of stoichiometric plasticity is known, stoichiometry of the food (or growth substrate) and its supply rate can be used to calculate both the growth rate of the organism and the excretion rate of various materials. These calculations are based on a simple budgeting approach. A primary constraint is that some knowledge of specific constraints is necessary to calculate demands (e.g., trees have high carbon demand for trunks and limbs, vertebrates have high calcium and P demands for bones).

This theory has led to numerous ecological predictions about how nutrient limitation is influenced by organisms, how nutrient limitation is propagated through food webs, and how nutrient recycling can feed back to primary producers and detritivores. Food web propagation and propagation through primary producers leads to predictions about how ecosystems can be influenced by stoichiometric constraints. For example, if a grazer is N-limited and the autotrophs the grazer eats are N-limited, the grazer will tend to intensify nutrient limitation. Stoichiometry explains, in part, why grazers select the food they do (e.g., Joern and Mole 2005).

Stoichiometry is commonly used to estimate nutrient limitation of primary producers (an extension of work by Redfield 1958). The elemental composition of phytoplankton reflects the limiting nutrient. For example, if ratios of C:P and N:P are greater in a particular ecological context than in plankton cells when they are under balanced growth (growth under nutrient replete conditions), P is likely the limiting nutrient. Tests of nutrient limitation (bioassays) support the idea that elemental composition of assemblages of primary producers can be used to ascertain the element in the environment that limits production (e.g., Dodds and Priscu 1990). The limiting nutrient or nutrients are reflected by tissue concentration because ratios in the tissues reflect the ratios of the supply rates of individual inorganic nutrients to primary producers (not their absolute concentrations, which may not be reflective of availability when turnover rates of inorganic nutrient pools are high).

An early version of prediction within stoichiometric theory is the resource ratio theory developed by MacArthur (1972) and refined and expanded by Tilman (1982). The model makes a number of predictions (see next paragraph) related to when one species will dominate in competition with another based on supply and use rates of two or more resources. The model specifically predicts which species will come to dominate under a defined set of conditions assuming equilibrium circumstances. A unique prediction is that when two or more resources are limiting, more than one species can coexist because of inherent trade-offs in the use of the resources (law of *Specialization*). Perhaps the strongest prediction from this theory is that if organisms have different competitive abilities, even under equilibrium conditions, Leibig's law of the minimum does not necessarily hold (see discussion in Chapter 1). Furthermore, adding spatial or temporal variability can lead to coexistence of multiple species with only two limiting nutrients (Fig. 5).

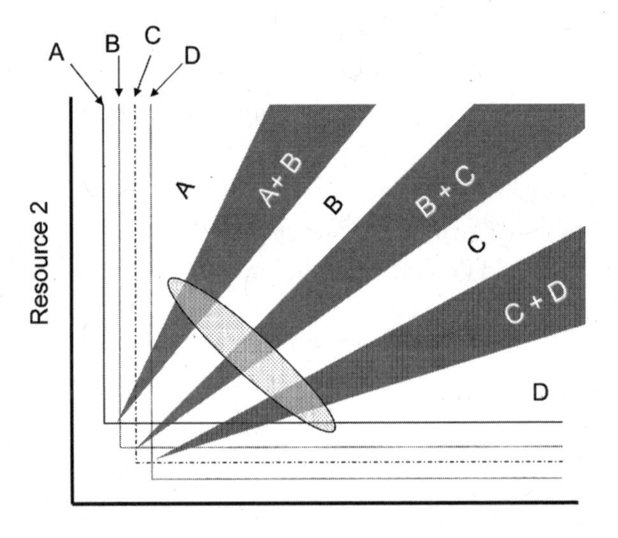

Figure 5. A graphical representation of Tilman's resource ratio theory for two resources (1 and 2), and four species (A-D). Nutrient concentrations increase from the lower left side of the graph. The zero growth isoclines for each species are denoted by the letter pointing to them. For example, A is a better competitor for resource 1 than all the other species because it does better than the others at low concentrations of resource 1 but a worse competitor for resource 2. Therefore, at high ratios of resource 2/ resource 1, A dominates. Species D dominates when resource 1 is plentiful relative to resource 2. At three ranges of resource ratios, two species can coexist. If spatial or temporal heterogeneity causes the ratios to vary (e.g., within the shaded ellipse), all species could coexist. [Drawn after Tilman (2007)].

Early tests of the theory using planktonic diatoms in culture generally supported the stoichiometric approach to resource competition (Tilman 1981). Miller et al. (2005) suggest that Tilman (1982) or subsequent researchers used resource ratio theory to make any of seven predictions. Then, Miller et al. (2005) surveyed 1333 ecological research papers that cited the resource ratio work by Tilman and found less than 1% of the papers claimed to test predictions of the theory. The predictions, followed by the total numbers of confirming tests per total studies that adequately tested the predictions are

1. Species that can survive at the lowest levels of a limiting resource will be the best competitor for that resource (8/13).
2. Species dominance varies with the ratio of the availabilities of two resources (13/16).
3. The number of coexisting species is less than or equal to the number of limiting resources in a homogenous environment (1/2).
4. Resource supply rates to an environment will affect whether competing species coexist and, if not, which species will competitively exclude the other (5/6).
5. Consumption rates of resources for two species will determine whether competing species coexist or, if not, which species will dominate competitively (2/2).
6. Trade-offs in resource use must occur for species to coexist along a gradient of ratios of the availabilities of two resources (2/3).
7. The highest diversity of competing species will occur at an intermediate ratio of the availabilities of two resources (0/0).

These tests suggest that the theory offers a moderate degree of ability to predict the competitive outcome between two species under natural conditions. The approach of using published literature to establish predictive ability could be problematic, because studies that report negative results may be less likely to be published (not controlled for in this review, although statistical procedures are available to test for this problem), so the results of this meta-analysis could indicate the upper level at which resource ratio theory is supported by empirical work. Presumably, if 1,333 ecological research papers cite Tilman's work, the theory has fairly broad importance, although not all authors necessarily indicate support of the work's predictions simply by citing it.

The theory provides predictability in highly controlled systems (e.g., a chemostat with two known competitors). Thus, conditions that approximate equilibrium where a few resources strongly control different members of the system are the most likely to be ecological systems in which prediction is possible.

Some Difficulties with Prediction The major lack of prediction in stoichiometric theory lies in the confounding influence of evolutionary adaptation. There is tremendous adaptive advantage in being less constrained by stoichiometry. For example, autotrophic plants are often not limited by carbon (assuming adequate light and water) and thus can make high-carbon structural parts. The vast majority of trees consist of carbon-rich organic carbon compounds. This ability to be plastic in stoichiometric composition (compared with green algae, the precursors to land plants, which do not need much structure to compete for light in aquatic environments) has led to tremendous adaptive success to the point where the carbon sequestered by land plants has global biogeochemical effects. Vertebrates have bones that are rich in P, mollusk shells are rich in calcium, and diatom frustules are composed of silicate. These are all examples of constraints that need to be considered when applying the theory of stoichiometry.

One of the earliest predictions of ecological stoichiometry made on a global scale is related to the work of A. C. Redfield (1958) on marine phytoplankton. He noted that phytoplankton from the open ocean had C:N:P:O stoichiometry similar to the ratios found dissolved in the marine waters. Based on his observations, he proposed three potential explanations: "(1) A coincidence dependent on the accidents of geochemical history; (2) adaptation on the part of the organisms; or (3) organic processes which tend in some way to control the proportions of these elements in the water." Redfield (1958, p. 210) dismisses the first explanation as improbable. The second is dismissed, because dissolved oxygen is determined by organisms (i.e., produced by photosynthesis). Redfield settles on the third explanation, control by biotic processes, for global marine biogeochemical stoichiometry. Redfield does not conclusively rule out the second explanation in favor of the third. Organisms on Earth could have evolved stoichiometry of biomolecules that best matched the availability of N and P in the environment. Some degree of evolutionary history is not escapable (Ricklefs 2007). For example, organisms require several nutrients to be in reducing form to enter into synthetic biochemical pathways (ammonium instead of nitrate, sulfide instead of sulfate, reduced iron) presumably, because this is the form these nutrients were in under the reducing conditions under which life evolved (evolutionary history has long-term influence on biochemical pathways). Organisms can use the oxidized forms, but must convert them to reduced forms before they enter cellular biochemical pathways. Organisms that were out of balance with availability (cellular stoichiometry not matching environmental stoichiometry) would have a competitive disadvantage, because they would have needs in excess of supply rates. Even given 50 years of biogeochemical knowledge accumulated since Redfield published his ideas, it is still not possible to substantiate explanation 2 or 3 at the expense of the other. The controversy over global control of biogeochemistry was raised again when the concept of Gaia was proposed (Volk 1998).

One way that stoichiometry has been used incorrectly is when the assumption is made that ratios of concentrations of dissolved inorganic nutrients are equivalent to the ratio of rates at which they are used. This assumption is problematic because pools of dissolved inorganic nutrients are dynamic, and concentration and flux rate are not necessarily the same. The most common example of this type of error in prediction is when attempts are made to use the ratio of dissolved inorganic N (the sum of nitrate, ammonium and nitrite concentrations) to soluble reactive P (thought by some to represent phosphate, but actually more P compounds are included in common chemical tests) to indicate limiting nutrients for primary production (Dodds 2003). This error in prediction could have large economic consequences as well as be misleading ecologically. A practical example of potential for misusing stoichiometry to make predictions based on ratios of dissolved inorganic nutrients comes from the eutrophication in the Gulf of Mexico that causes problems of hypoxia. If it is incorrectly assumed that inorganic P could be used to indicate P limitation, inappropriate management actions could be implemented based on misunderstanding of how nutrient dynamics relate to stoichiometry (Dodds 2006).

Food web interactions may complicate ecological stoichiometry (Hall et al. 2007). Predation can promote nutrient limitation of herbivores, so trophic cascades can interact with stoichiometric controls. Given that the factors driving trophic cascades are difficult to predict (see Chapter 3 and Hall et al. 2007), food web effects translate into difficulties making predictions with stoichiometric theory.

Ecological Diffusion

Materials and organisms move across the environment leading to predictable ecological patterns. When particles (e.g., organisms, molecules) move passively, they will spread from high to low concentration. Basing diffusion purely on Brownian motion is too restrictive, because diffusion mathematics also applies to advective or transport diffusion as long as the material being moved cannot move faster than the transport. Advective transport applies to any situation in which there is turbulence (most open water aquatic environments, the atmosphere) or where directional abiotic transport mechanisms operate at more rapid rates than molecular diffusion (streams, water flowing through soils, wind currents). Gravity often moves larger particles more rapidly than molecular diffusion rates and drives advective transport. Thus, in this theory, diffusion is a general term for particles or chemicals moving from high to low concentration via physical transport whether by molecular diffusion controlled by Brownian motion or advective transport aided by bulk movements of the media containing the particles. The diffusion framework provides a macroscopic view of aggregate random behavior. The basic mathematics of diffusion can be modified for many ecological problems (Okubo and Levin 2001) and used to form null models of how organisms or materials spread in the environment.

Laws and Theories Most Applicable Laws from physics and chemistry. System openness. Scaling, Population and resource heterogeneity.

Strong Patterns

- Particles substantially larger than the media they are in tend to move with gravity.
- Advective transport rates exceed molecular transport rate and alter diffusion patterns (e.g., transport associated with water flow in streams).
- Larger differences in concentration lead to greater rates of flux. Flux rates are greater over shorter distances. These relationships are formalized by Fick's law in which the diffusive flux,

$$dC/dt = D(C_1 - C_2)/(X_1 - X_2) \tag{14}$$

is determined by C, the concentration at point 1 or 2, X gives the spatial location of the two points, and D is the diffusion rate coefficient. When energy for movement of particles comes from outside, the process is called passive diffusion; it is called active diffusion when the particles provide energy.

- Organisms tend to evolve dispersal mechanisms that accentuate their rates of diffusion.
- Organisms have sensory adaptations that allow them to move from resource-poor areas to resource-rich areas more rapidly than by random diffusion. This behavior causes them to work against diffusion and maintain a high concentration of organisms in a small area.

Some Predictions Turbulence overrides molecular diffusion and causes patterns similar to those observed for molecular diffusion, but Fick's law does not necessarily hold, because the timescales of movement of individual particles are similar to those where observation occurs. In contrast, molecular diffusion is the result of many molecules moving rapidly but changing direction quickly, so each individual movement is small scale (Okubo and Levin 2001).

Physical constraints alter advective diffusive movements. For example, a plume of chemicals released in a river will move in a downstream direction as well as spread laterally. An example of this type of constraint is the downstream movement of sewage from a treatment plant releasing into a river. Eventually, all the sewage is mixed with the river, but initially there are strong concentration gradients from inside to outside the plume. Salmon from the ocean, homing in on the natal stream where they will reproduce, use the physical properties of diffusion gradients (follow a plume of odor to its concentrated source). Another example is the movement of airborne propagules in a directional wind. The air shed downwind from a reproductive wind-fertilized plant for pollen release is a diffusion plume driven by unidirectional advection. Immediately downwind the pollen delivery rates are high in a very narrow band. Farther downwind, the pollen cloud spreads laterally and the delivery rates are not as great, but the area covered by the pollen is greater.

One prediction of diffusion is the spread of an organism that is expanding its range. The null model would be the expectations generated by diffusion theory. Deviations from that expectation may yield clues to avenues of dispersal specific to those organisms.

A prediction based in part on potential energy and ecological diffusion is that edges tend to be some of the most productive habitats. Edges are often productive across spatial scales from microbes to large animals, because edges are zones with concentration gradients and flux rates are greater across such concentration gradients. There are other examples of edges being productive (e.g., allowing organisms to access food on one side of the edge while receiving protection from predators on the other side of the edge), but concentration gradients also can stimulate productivity.

An example of consequences of diffusion of global importance comes from the microbial world. The zone of contact between oxic and anoxic habitats is an active zone for two reasons (1) sharp gradients (short distances) lead to rapid diffusion rates and (2) reduced compounds diffusing from reducing (anoxic) areas into oxidized areas have potential energy as do oxidized compounds diffusing from oxidized areas into anoxic zones (Dodds 2002). These zones are areas of high biochemical metabolic diversity (including bacteria that oxidize sulfur, ammonium, methane, reduced iron, and reduced manganese as well as anoxic bacteria that use nitrate, sulfate, oxidized iron, and oxidized manganese as electron acceptors to oxidize organic carbon). Any place with such zones exhibits these characteristics including, but not limited to, anoxic zones in decaying leaf materials, reduced sediments with oxygenated water over the sediments, the zone between the anoxic hypolimnion at the bottom of a stratified lake and the warmer oxic water floating on top, an aggregate of organic material in moist soil, a fecal pellet in oxic water, and hydrothermal hot springs at the bottom of the ocean.

Diffusion forms the neutral model of movement of organisms and materials. Organisms dispersing against diffusive flux, or concentrating themselves, requires energy and probably has some adaptive significance. Many aspects of ecology cannot be predicted via diffusion theory, but this does not make the theory useless.

Diffusion theory predicts a pattern observed when organism distribution is dominated by abiotic transport.

Some Difficulties with Prediction Movements of animals can reverse or accentuate ecological diffusion. An example of an organism accentuating diffusion is the hippopotamus, which grazes on land at night and finds shelter in the water by day. Hippopotami increase the rate of nutrient movement from land to water and inhabit an area near the edge of where resources they depend on (terrestrial plant material) are available, and the area where they seek shelter. Similar examples include beavers or any other organisms that spend more time lower in a watershed but forage higher up.

The whole concept of food web subsidies (Polis et al. 1997, discussed in more detail in Chapter 3) could be considered an ecological result opposite that expected by ecological diffusion theory (organisms bring materials against the diffusion gradient because they move across the gradients for other reasons). Although nutrients tend to flow downhill and downstream, in some cases this can be reversed when animals move materials against the physical and chemical laws of passive diffusion. A well-studied case of nutrients being transported against physical gradients occurs when anadromous fishes swim upstream to spawn and die. Colony nesting birds that feed in aquatic habitats move nutrients from the water and excrete them under their roosts.

Animals maintain dense aggregations (e.g., herds, flocks, schools), because the benefits to survival select for behaviors that keep high concentrations of individuals in close proximity. These types of movement depend on evolved capabilities of animals, which may work contrary to trends imparted by minimizing potential energy and the process of ecological diffusion.

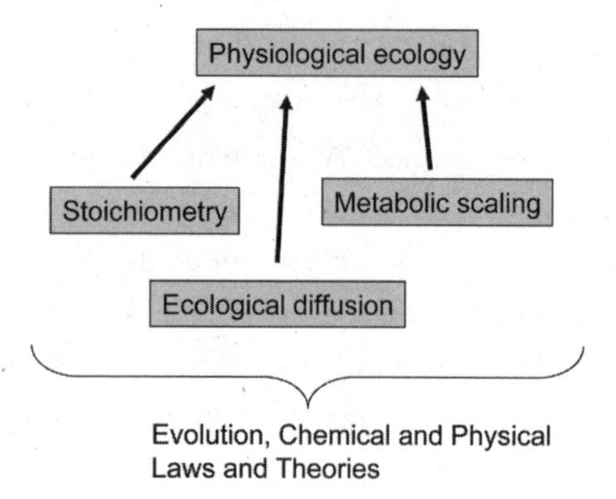

Figure 6. Some theories of physiological ecology covered in this book.

PHYSIOLOGICAL ECOLOGY

The physiology of organisms is closely linked to constraints imposed by chemical and physical laws and to the evolutionary response to those constraints and ecological context. Metabolic scaling theory is a major theory underlying physiological ecology (Fig. 6) that explicitly considers the role of scale in physiological constraints, and ecological diffusion is important as well. Stoichiometry and diffusion, described in the last section, also constrain physiology.

Metabolic Scaling

Brown et al. (2004, p. 1771) state that "metabolism provides a basis for using first principles of physics, chemistry, and biology to link the biology of individual organisms to the ecology of populations, communities, and ecosystems." The rate at which organisms metabolize depends on temperature and body size. Dependencies on central chemical and physical processes form the basis of the theory. Organism size has been used as a predictor of absolute and mass-specific metabolic rates for over a half century (Peters 1991). Brown et al. (2004) reviewed several simple metabolic scaling equations that can be used to predict metabolic rate across a huge range of orders of magnitude of animal, microbe, and plant size.

Huxley (1932) mentioned that most variation in characteristics of organisms as related to body size can be described by power functions. Basal metabolic rate was shown proportional to body mass early in the 20th century and takes the form of a power curve (reviewed in Kleiber 1932). This relationship was proposed to be

$$I = I_0 M^b \tag{15}$$

where $b = 3/4$, M = mass, I the individual-specific metabolic rate, and I_0 a normalization constant (intercept). The value of $b = 3/4$ has been referred to as "Kleiber's law" or "the 3/4 power law" (Glazier 2006). This relationship has been used to correct metabolic rate measurement experiments for different sized animals. Many other characteristics of organisms, such as the size of individual organs relative to body mass, take the form of a power law (Huxley 1932). Peters (1983) reviewed published allometric relationships that have been established for animals as a function of body mass including metabolic rates, physiological rates (e.g., blood circulation), velocity of locomotion, ingestion rates, mass of offspring, time to reproduction, nutritional and water requirements, and home ranges. Brown et al. (2004) reviewed how temperature correction of metabolic rate can be used to bolster the accuracy of prediction of metabolic theory. The theory of metabolic scaling is generally characterized by highly significant relationships that occur across many orders of magnitude of scale; mechanisms for the patterns have remained more elusive and controversial. The theory of metabolic scaling has given rise to a good part of a subdiscipline of ecology, macroecology.

Laws and Theories Most Applicable Laws from physics and chemistry. Recycling rates, Energy requirement, Maximum metabolic rates, Scaling, Temperature optima.

Strong Patterns

- Organisms have a maximum efficiency and basal metabolic rate that depends on fundamental evolved characteristics (e.g., single or multicellular, thermal homeostasis).

Some Predictions Body mass of organisms varies over 15 orders of magnitude. There are highly significant statistical relationships between body mass and physiological characteristics, such as resting metabolic rate, biomass production rate, and carbon turnover rate across these scales (Brown et al. 2004). For example, body mass can be used to predict metabolic rate over a wide range of organism sizes (Fig. 7). Body mass is positively correlated with ecological properties of species, such as animal home range size (Jetz et al. 2004).

Metabolic rates can be corrected for temperature and body size in physiological experiments. Such correction allows for comparisons to be made across organisms and groups of organisms (e.g., endotherms have greater metabolic rates per mass than ecotherms). That this relationship occurs across 17 orders of magnitude of body size and 15 orders of magnitude of metabolism rate is astonishing. A range of animal size is a result of the law of diversification; the correction for temperature is a consequence of the law that chemical reaction rates depend on temperature.

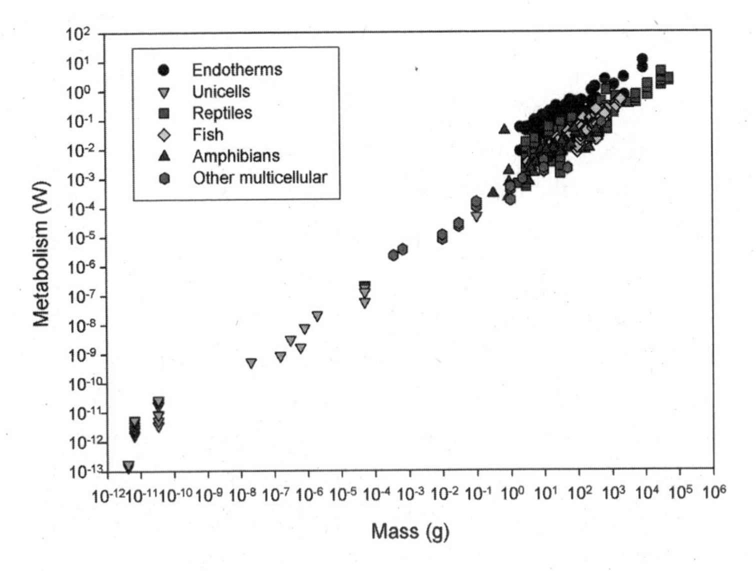

Figure 7. Metabolic rate as a function of body mass for animals and unicellular organisms. (Data from Gillooly et al. 2001).

Essentially, the upper edge of the line of points in Figure 7 is dictated by maximum power efficiencies as constrained by chemical and physical laws. It is probably not physically possible for an organism to evolve to inhabit the upper left triangle of open space on the plot. The lower right triangle of open space in Figure 7 is constrained by evolution not allowing continued success of less efficient organisms.

This theory can also provide a foundation for ecosystem and food web energetics because it allows prediction of the power output of different sized organisms. Again, the question of adequate degree of predictability with this relationship arises. At any mass, there is a one to three orders of magnitude variance in metabolism, more so with larger multicellular organisms. Constraints imposed by evolutionary history explain some of the variance seen with larger organisms, so endotherms have more power output per unit mass than fish or reptiles.

Some Difficulties with Prediction Although there are significant power relationships across scales, what the mechanisms are and what the slope of the relationship should be is less clear. The "¾ power" law has been used by many researchers, but the original exponent was thought to be ⅔. The ⅔ exponent is based on the observation that surface area is related to volume by approximately a ⅔ exponent and that metabolism should be controlled by rate of exchange across biological surfaces. The "¾ power law" has long been known to have exceptions (Kleiber 1932). An extensive analysis of metabolic rate of marine organisms with pelagic larvae or planktonic life stages suggests that an exponent of 0.95 rather than ¾ better fits the observed distributions (Glazier 2006). Analysis of data sets for mammals and birds led Dodds et al. (2001) to claim that there is no statistical evidence for an exponent of ¾ instead of ⅔.

West et al. (1999) noted that the fractal distribution networks (circulatory systems) of animals lead to a prediction that the exponent b should equal ¾. The fractal distribution network is a very attractive argument for mechanism behind metabolic scaling relationships because it is based on fundamental physical principles. Even if this exponent holds for many larger animals (and it is a controversial explanation, Cyr and Walker 2004), it might not apply to organisms that have no circulatory system (Glazier 2006). Unicellular Eukarya, Bacteria, and Archaea make up much of the cellular biomass on Earth (given the discovery of deep subsurface microbes below the continents and the sea floor suggesting a greater volume of archaeal plus bacterial protoplasm in the world than eukaryotic), and in my treatment, a general ecological law or theory should encompass most organisms. The patterns of metabolic scaling apply across most organisms, but the specific prediction of the exponent of the power law based on the mechanisms proposed by West et al. (1999) may not.

Examining the data in Figure 7 reveals that across 17 orders of magnitude increase of magnitude of body size, resting metabolism increases by 15 orders of magnitude. This mismatch between rates of increase implies a cost to larger cells, multicellularity, and relatively large body size. Exchange of materials occurs across surfaces and cellular metabolism in cell volumes, but organisms have a wide variety of ways to increase surface area to volume (e.g., lungs, spines, and projections), so exchange rate limitations alone do not necessarily explain the apparent cost to animal size.

In addition, the power law applies to resting metabolic rate, but organisms spend widely different amounts of time close to resting rates. The conditions under which they will not rest are difficult to predict. Physiological adaptations (e.g., endothermy) allow some organisms to rapidly change from resting metabolism to substantially greater rates of energy output. Flying animals (birds and bats) have greater maximum metabolic rates than running mammals (Schmidt-Nielsen 1984).

Metabolic rates of plants also vary with size. West et al. (1997) claim that scaling is a function of space filling models of vascular systems when the terminal tubes are the same size and energy dissipation is minimized, and this leads to a ¾ power law as constrained by plant vascular system architecture. In a subsequent review, Enquist (2002) makes a variety of predictions on plant architecture and vascular properties based on mass relationships to vascular systems and finds strong support for many of the derived allometric relationships. Reich et al. (2006) found data from 500 terrestrial plants across 43 species indicating that whole-plant respiration rates most closely relate to whole-plant N content (this approach could partially control for N limitation). They did not find the plants to follow the "¾" rule.

Scaling of plant metabolism has also met with prediction difficulties. A group of 41 ecologists analyzed data on tropical forests from 10 old-growth stands, with census data on 367 ha and more than 1.7 million trees, to test the ability of allometric scaling to predict diameter growth rate and mortality. Their analyses suggested that growth was not related to tree size (Muller-Landau et al. 2006). These scientists tested the assumptions and predictions of plant allometry published by West et al. (1997, 1999), including the expectations that a plant's biomass growth rate scales with $mass^{3/4}$ and an individual's mortality rate scales with $mass^{-1/4}$. Neither prediction was supported.

The case has been made by Martínez del Rio (2008) that the body of research based on metabolic scaling is more reasonably thought of as a series of models rather than a theory. This case has merit because the foundational mechanisms, particularly behind what the scaling exponent should be, are not well established. Even the fundamental basis of thermodynamic scaling used in metabolic theory, the Arrhenius equation, is regarded by physicists as an approximation (Martínez del Rio 2008). These arguments have been echoed by O'Connor et al. (2007) who identify optimal material transport as a poorly developed mechanistic base for metabolic theory. The argument presented by Martínez del Rio could apply to all the theories presented in this book; the basic approach of insisting that science must be based on theories from first principles could actually hinder scientific progress. Allen and Gillooly (2007) suggest that deep philosophical differences, based on the issue of the existence of laws in biology, drive the debate over metabolic theory. In the case of metabolic theory, general predictions can be made based on the empirical patterns at least, and there are still mechanistic bases behind some of those predictions (e.g. dependence of metabolic rates on temperatures, Allen and Gillooly 2007). Scientists may nail down some mechanisms related to energy and material distribution in organisms in the future, but these basic principles will need to be constrained based upon the organisms considered (e.g. plants, animals, single celled microbes).

POPULATIONS

Populations within the same species have been studied extensively by ecologists since inception of the field. Applying constraints, such as spatial structure, allows for broader predictions, such as those in the theory of metapopulations (Fig. 8).

Population Biology

Population biology has a strong mathematical basis in ecology and a long history of building a theoretical foundation. Turchin (2001), Berryman (2003), and Ginzburg and Colyvan (2004) have proposed laws that could build a general theory of population biology. Here, laws proposed in Chapter 1 form the basis of the theories.

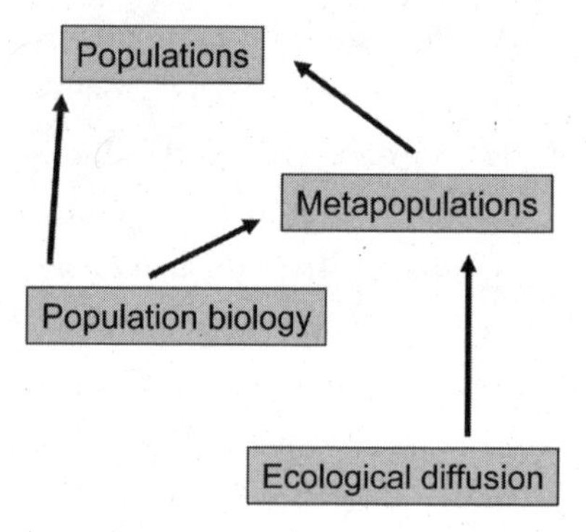

Figure 8. Some theories predicting properties of populations of organisms discussed in this book.

Laws and Theories Most Applicable System openness. Conservation of individuals. All organisms die. Exponential growth. Limits to growth. Population stability not determinant. Extinction probability. Biotic/abiotic interaction. Population heterogeneity. Species interact. Scaling.

Strong Patterns

- Sigmoidal population trends occur when abundance is a function of rate of resource supply.
- Many populations persist over evolutionary time; thus, very unstable population dynamics are not likely for these populations.
- Density dependence is a widespread feature of population dynamics (Brook and Bradshaw 2006).

Some Predictions The concept of exponential growth can be useful in situations where constraints on population growth are relaxed. When an invasive species is initially successful in a habitat or disturbance opens up habitat with minimal resource limitations, initial growth of the population is approximately exponential. The law of exponential growth combined with the finite resource base on Earth leads to substantial concern over what will control the super-exponentially growing human population. The law of exponential growth and population limitations led Darwin to propose natural selection as one basis of the theory of evolution.

The exponential growth law describes growth of single species when released from food and predation constraints, and as food becomes limiting, the logistic growth curve fits relatively well (Fig. 9). Single species in culture are an artificial situation, but following disturbances, many species may be released from resource and predation limitation.

As a population reaches a point where limiting factors take effect, it can follow one of three trajectories (1) come to a steady equilibrium, (2) oscillate cyclically, or (3) enter a chaotic state (Coulson and Godfray 2007). These functional responses demonstrate that the full range of possibilities occur (i.e., that with limitations, populations will be stable or unstable, and if they are stable, they may oscillate or not). Time lags inherent in natural populations allow these instabilities to occur.

One of the primary constraints in simple population models is that all individuals are equally able to reproduce (have an equal probability of reproducing). Life history matrices, representing age structure in communities, relax this constraint and allow for parceling out the contributions of various life stages to survival and reproductive output (Coulson and Godfray 2007). Including life history matrices leads to rich and varied population behavior including analyses of survival as a function

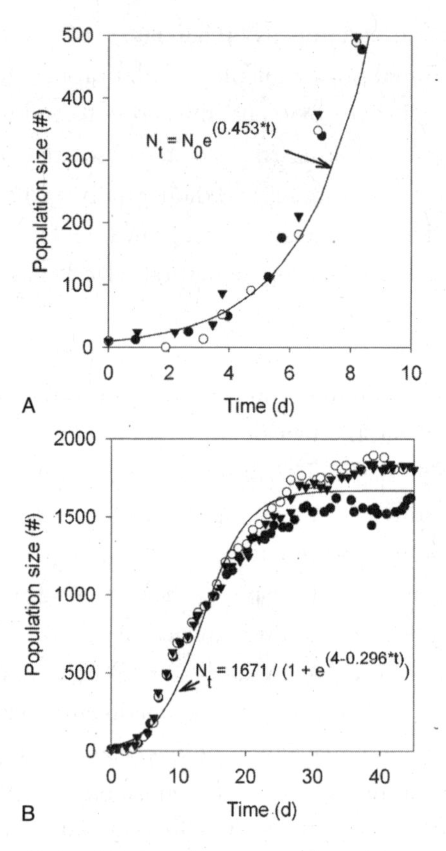

Figure 9. Population growth of the rotifer *Brachionus plicatilis.* Three cultures were started with 10 individuals each in 1.5 mL of growth medium with an algal culture. All individuals were transferred to fresh medium with the same algal density daily. In (A) the first 10 days of growth are fit to an exponential curve and in (B) 45 days of growth are fit to the logistic growth curve. (Data from Yoshinaga et al. 2001).

of life history stage to determine controls on growth or decline of populations.

Budgeting of population numbers allows for construction of population models. Population models are useful because they allow determination of probability of survival and reproduction at each life stage. Managers of populations of plants and animals can use this information, and population models can also be employed to determine the demographic parameters and sensitivities where species of conservation concern are most vulnerable to decreases in survival or reproductive output.

Population viability analyses predict the probability a population will go extinct given specific life history information. An assessment of the ability of population viability analysis to predict extinction was conducted using 21 long-term population records. Half the data for each population were used to predict population trends, and the other half were used to compare predictions to actual population numbers. These data suggested that population viability analyses were good at predicting the probability of extinction of individual populations (Brook et al. 2000), although the paper is controversial (e.g., Coulson et al. 2001).

Brook and Bradshaw (2006) tested long-term abundance time series of 1,198 species and found strong evidence for density dependence in population dynamics. Dependence was not taxon specific. Intensity of density dependence varied among species, but the longer the time series, the more likely the detection of density dependence. Thus, density dependence is a core concept and prediction of population theory.

Some Difficulties with Predictions One of the most intriguing aspects of population biology is what cannot be predicted; chaotic behavior of population models occurs under certain conditions. Robert

May (1976) did some of the earlier work on the complex patterns that could arise from some mathematical equations for density-dependent population growth. He found that with rapid enough population growth rates, small deviations in initial conditions could lead to different system states. May's work, in part, gave rise to a new branch of the field of mathematics, chaos theory.

With small populations, individual-based models may be more appropriate than entire population models. In contrast, when large populations occur (such as bacterial populations in a lake that reach hundreds of thousands per milliliter) and resources suddenly become very abundant, the approximation of exponential population growth could allow very accurate prediction. Even if individual population models work well in a specific ecological setting, applying multiple population models (e.g., population models to community ecology) has difficulties.

Lotka-Volterra predator-prey models are nonlinear differential equations that can describe the interaction between predator and prey (Lotka 1925, Volterra 1926). They predict stability under some conditions and population oscillations under others. An analysis of 634 population studies of animals indicated that most populations were steady (Sibly et al. 2007). Unfortunately, Lotka-Volterra equations are not easy to use to predict exact numbers of individuals in ecological populations because so many simplifying assumptions must be made. They do provide a good description of potential modes of population behavior (patterns to look for in populations).

If Lotka-Volterra models are modified for mutualism, some problems start to surface; additional constraints must be applied to the model to allow for stability. Specifically, if two populations interact as mutualists, self-regulation must be greater than the benefit of the interaction to allow for stability (Post et al. 1985). There is not a strong evolutionary argument for a species to evolve self-regulation in order to not be too successful. If most of a population is self-limiting but a few cheaters arise, the cheaters will overtake the rest of the organisms that are cooperating by self-regulating. This nonregulation could be selected for, even if it ultimately dooms the species to extinction. However, mutualisms are common and essential for survival of most species in the real world. Gut microbes of ruminants and termites (and most other animals), coral reefs, pollinators and flowers, mycorrhizae and most plants, lichens and endosymbionts, and numerous other species provide examples of mutualisms that are stable over ecological and evolutionary time. Mutualism may even be more common among microbes than macro-organisms (Ross-Gillespie et al. 2007). Mutualism is an area of fertile research and numerous solutions to how cooperation evolves and persists have been proposed (Nowak 2006), but given the contingencies, predicting where cooperation will occur is not part of the theory of population biology as presented here.

Metapopulations

The simple observation that most populations of organisms are not closed but there is variable amount of gene flow among different populations that leads to metapopulation theory. The theory has been refined and explored by landscape ecologists and has implications for conservation biology as well as the spread of invasive species, diseases, and novel genetic materials.

Metapopulation theory is based on the ideas that individual populations interact and interaction has consequences for population dynamics. Thus, interactions of multiple connected populations can be considered one of the hierarchical levels of biological organization (Hanski and Gaggiotti 2004). The law of heterogeneity dictates that populations will not be found everywhere across a spatial area. Dispersal is the "glue" that holds metapopulations together (With 2004).

Laws and Theories Most Applicable Laws from physics and chemistry. System openness. All organisms die. Conservation of individuals. Exponential growth. Extinction probability. Population and resource heterogeneity. Scaling.

Strong Patterns

- Levins' model (equation 10, Chapter 1), described in law of population and resource heterogeneity.
- Levins' model also leads to Levins' rule "A necessary and sufficient condition for metapopulation survival is that the remaining number of habitat patches following a reduction in patch number exceeds the number of empty but suitable patches prior to patch destruction" (Hanski et al. 1996, p. 528).

Some Predictions Levins' rule leads to a prediction of the minimum requirement for remaining habitat patches (but this is a minimum). This rule provides an important prediction for conservation biology. Managers must know how small a population can be without risk of extinction before attempting to conserve the population. Although putting a solid number on how small the population will be relative to the probability of extinction is somewhat difficult, the idea that there is some minimum viable population is an important one.

Initial models predicted by Levins' rule and other early work assumed that extinction was equally likely in individual patches.

Space-structured dispersal frequency and continuous population dynamics were introduced, but the model assumed equal likelihood of extinction in patches of similar size. Harrison (1991) suggested that a more realistic view of local extinction probability would make the models more predictive. She suggested there is support from empirical data for application of metapopulation theory where there are (1) source-sink dynamics with extinction resistant populations, (2) patchy populations with high interpatch dispersal rates that are overall resistant to extinction, and (3) nonequilibrium conditions where the population is decreasing and continued extinction of local populations leads to overall population declines.

Some Difficulties with Prediction Harrison (1991) provided a detailed empirical review of metapopulation theory with respect to extinction and dispersal. She claimed that there is little empirical evidence for dynamics of natural populations being well described by Levins' and related models. Part of the lack of prediction may be related to restrictive model assumptions. For example with respect to constant probability of extinction across patches, some disturbances (e.g., an unusual freeze) may cause many or most populations to go extinct simultaneously, but others (e.g., tornados) may be more spatially limited.

Baguette (2004) makes a more pointed criticism of what he calls classical metapopulation theory. He maintains that the predictions of classical metapopulation theory are often not applied to conservation management and minimum population viability analyses. He argues that tests of metapopulation theory are most often done on very small populations at the edge of ranges of distribution with declining populations. These populations are far from the equilibrium assumed in the original classical models. Baguette

(2004) does not discard metapopulation theory, but points out that the constraints of the model make it difficult to apply to many populations and metapopulation models should be used with care based on the specific characteristics of the populations being studied. A more recent review (Driscoll 2008) mirrors these criticisms, suggesting that definition of a metapopulation is not consistent across studies and current models are not often useful to managers, except possibly to rank various management strategies.

Gotelli and Taylor (1999) tested the predictive ability of a generalized version of Levins' rule on a 10-year time series of spatially explicit fish collections from the Cimarron River. Their analysis suggested that the simple model failed to predict the behavior of most fish populations. A model that incorporated spatial variability in colonization and extinction probabilities would be required to more accurately portray population dynamics in this river. Thus, the general model can be improved by introducing more constraints (as is usually the case).

Another problem with metapopulation theory is how to deal with cosmopolitan species, particularly those that reproduce asexually. Any species that is found almost everywhere will not easily go extinct because replacement individuals are continuously available. Most protozoa have worldwide distribution (Finlay 2002), and bacteria probably do as well. Bacteria can exchange genetic information via conjugation, plasmids and viruses. The exchange of bacterial genetic material can be specific or genes can move among bacteria that have a widely divergent taxonomy. How bacteria even maintain species in the face of this gene flow is not well understood (e.g., Andersson and Banfield 2008). For example, genes that allow resistance to human-synthesized antibiotics are passed on plasmids. This method of genetic transfer broadcasts the genes widely; antibiotic resistant species are cosmopolitan, even in habitats where human synthesized antibiotics are never found in detectable amounts. The concept of metapopulation applied to bacteria and protozoa is tenuous at best.

Species that are very patch-specific (e.g., insect larvae that rely on a single species of moderately abundant plant) may be well described by patch-dynamic approaches relative to species that can easily traverse large expanses between patches (e.g., some birds). How are populations even defined given variable rates of gene flow among populations? How does gene flow proceed? These are issues of population genetics that are yet to be solved or directly addressed by a general theory of metapopulations. On the bright side, the more widespread availability of genetic techniques makes characterizing gene flow rates, and thus characterization of metapopulations, substantially more accessible to population ecologists. This developing field of landscape genetics (Holderegger and Wagner 2008) allows better characterization of what is a population and how well populations are linked, so a predictive mechanistic basis for metapopulation ecology is getting within reach.

COMMUNITIES

I present a general theory of community structure and constraints to allow more specific theories. Theories from population biology describe behaviors of groups of related organisms. Constraints on diffusion of organisms and their probability of survival upon reaching isolated habitats help build a theory of island biogeography that has been extended to a general theory

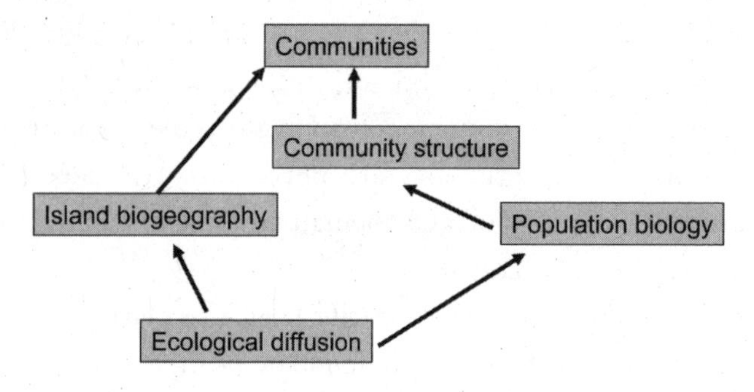

Figure 10. Relationships among some theories discussed in this book related to community ecology.

of species accumulation with greater area. These theories are part of the framework of community ecology (Fig. 10).

Community Structure

The laws proposed in Chapter 1 allow formalization of a theory of community structure. A general theory can be constructed by combining these laws with some strong patterns (observations with too many exceptions to be classified as laws that can be considered probabilistic laws); a theory of community structure is based first on a definition of community.

My working definition of a community is all the species of organisms found in a defined area over ecological time. The definition is similar to that provided by Holyoak et al. (2005, p. 8) who define a community as "the individuals of all species that potentially interact within a single patch or local area of habitat." If the law that all species interact with others is accepted, then all species that are in an area interact, at least indirectly, with all others, making the two definitions similar. My definition recognizes the subjective nature of defining a community by allowing people to classify the area of interest. My definition does not allow a community to be limited to a subset of the species found in a specific area unless such limitation does not alter fundamental properties of community structure (that is viewed here as too strict of a constraint for a general theory of community structure). The constraint of the term community not referring to a defined subset of species in an area is avoided by using the term *assemblage*. Some "natural" boundaries may define areas of study, such as lakes, islands, or mountains where there are relatively clear biologically relevant boundaries; but in other areas, the spatial and temporal boundaries of the community are more subjective. The laws on system openness and species interaction dictate that closed communities are a simplification for the convenience of ecologists. Holyoak et al. (2005) provide a more complete discussion of the various possible definitions of community.

Laws and Theories Most Applicable Theory of population biology. Theory of evolution. System openness. Biotic/abiotic interaction. Theory of evolution. Evolution affects ecology. Specialization. Scaling. Species interact. All interaction signs possible. Diversity of interspecific interaction. Variance of interspecific interaction. Competitive exclusion. Dominance of *Homo sapiens*. Linkage of interactions. Nonpropagation of interaction chain. Heterogeneity increases diversity. Diversity positively correlated with area.

Strong Patterns

- Interaction webs among species are structured by the fact that the primary energy flux is dominated by primary producers or heterotrophic decomposition of allochthonous organic materials.

- Terrestrial habitats are dominated by plants where adequate moisture is present, and phytoplanktons are the dominant producers in much of the ocean and large lakes.
- Microbes and plants will dominate primary energy inputs; animals are primary consumers or predators.
- Free-living microbes or those in animal guts will dominate detritivory.
- Several trophic levels exist above the primary producers or detritivores, but these levels are not necessarily distinct.
- Some omnivores will be found that eat at several trophic levels.
- A habitat that has many species removed will reassemble and there will be a temporal pattern of increase in the number of species, but the increase will not be sustained indefinitely (succession).

Some Predictions This theory predicts the existence of food webs. Because organisms have an energy requirement and many have adapted to eat other organisms, food webs are a central part of community structure. In terrestrial habitats and many aquatic habitats (Dodds and Cole 2007) connected to terrestrial habitats by drainage networks, terrestrial plants ultimately supply the primary energy source.

Communities are not fully connected because there are some zero interactions (there is interaction structure), and there will be organisms on the same trophic level that tend to compete with each other. The idea that communities are not fully connected is strengthened by theoretical analysis, indicating that communities that are randomly connected and fully connected are unlikely to be stable (May et al. 2007). Because many communities persist over a substantial length of time, some degree of community stability is likely. Organisms will have distinct ranges or boundaries to their range of distribution, and these boundaries will govern species dynamics, such as changes in ranges and how they interact with other species (Holt and Keitt 2005). Other strong predictions about community structure are difficult, but numerous patterns that may apply include trophic cascades, maximum diversity at intermediate disturbance related to community interactive processes, and keystone predators. However, as will be discussed later, these patterns can occur, but are not useful for general prediction.

The fundamental vegetative base (biome) of communities can be predicted fairly accurately if rainfall and temperature (latitude or altitude) are known (Holdridge 1947). This prediction is based mainly on empirical observations. Some mechanisms (e.g., conditions that are too harsh above timberline or too far from the equator to support trees) are involved in these predictions. Because vegetation provides so much of the structure for animal assemblages, the ability to predict vegetation is a substantial predictive success.

Existence of functional groups can be predicted. At the crudest level of prediction, primary producers and consumers (trophic levels) are predicted to occur in all communities. Detritivores also occur in all communities. Terrestrial habitats with flowering plants have pollinators, plants have mycorrhizae, all animals have symbiotic gut microbes, and planktonic communities have filter-feeding consumers.

With application of additional constraints, subtheories can make more specific predictions. For example, the theory of island biogeography (discussed in the next section) predicts the relative species diversity of habitats as a function of habitat size and distance from source of colonists.

Some Difficulties with Prediction As a practical matter, all organisms in an area cannot be considered. Some are so rare they are likely unimportant, others are simply not described (e.g., most of the Bacteria and Archaea). However, my definition proposed at the start of this section on community theory does not allow for subsets of a community (e.g., a "community" of birds or a "competition community" as modeled by theoreticians) to be truly considered a community without proving that other interactions are not essential to community properties. For example, a group of birds could be potential competitors, but if their predators always keep populations low, then competition is not necessarily important. The difficulties with defining the community as a unit of ecological study have been discussed in detail in the ecological literature (e.g., Holyoak et al. 2005, Jax 2006). The concept of a "metacommunity" (Wilson 1992, Leibold et al. 2004) further blurs the boundaries and connections of communities.

Food webs are tremendously complex. Take for example the marine food web illustrated in Figure 11. The primary food sources, phytoplankton and detritus are actually substantially more complex than reported in the publication that analyzed the food web (Link 2002). Detritus no doubt is a complex community, including bacteria, protozoa, and small animals. Phytoplankton is composed of numerous species, some much more resistant to grazing than others. Many of these species that are depicted feed at several trophic levels. Despite the complexity of this web, the strength of individual interactions and how they change over time is not depicted.

The exact form of food webs is difficult to predict because omnivory blurs trophic levels. Other feeding relationships, such as

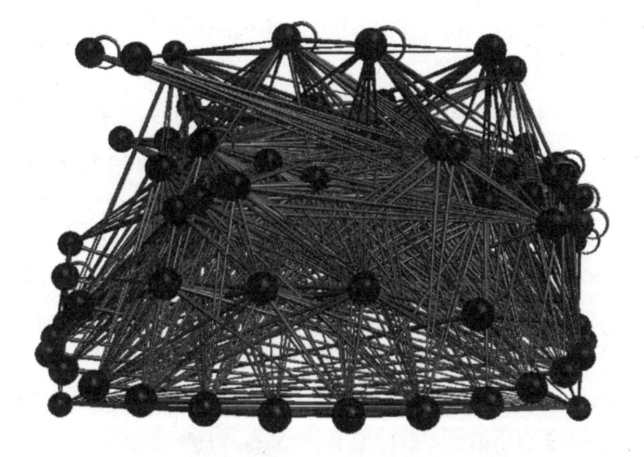

Figure 11. A three-dimensional graphical depiction of the food web assembled from sampling the northeast U.S. continental shelf from Cape Hatteras, North Carolina to the Gulf of Maine (Link 2002). Primary food sources, phytoplankton and detritus, are represented by the two small balls at the very bottom left and right of the figure. The lower ring of organisms is the primary consumer animals. In general, organisms at higher trophic levels are depicted higher up in the diagram. Lines are drawn from food source (narrow part of the line) to predator (thicker part of the line). Cannibalism is depicted by a link from a species node to itself. Image produced with FoodWeb3D, written by R. J. Williams and provided by the Pacific Ecoinformatics and Computational Ecology Lab (www.foodwebs.org, Yoon et al. 2004).

parasitism and mutualistic gut microbes that aid in digestion, also make it difficult to make strong predictions about the structure of food webs. Some investigators have attempted to make generalizations using published food webs (e.g., scale invariance in community structure). Interaction strength is a key feature in food web structure (Krause et al. 2003); unfortunately, few food webs have defined interaction strengths.

Claims of generality across food webs have commonly been met with skepticism (e.g., Havens 1992, 1993, Martinez 1993). Problems occur when food webs are not well resolved (all species and links are not identified, Martinez 1991), which is the case with most food webs. The microbial component of many food webs remains poorly resolved because most microbes have not been identified and their ecological roles are not well established. Processes that control the length of food chains have been debated for decades (Pimm 1991, Rosenzweig 1995). Analysis of described food webs suggests the length of food webs is not predicted by efficiency of energy transfer and basal productivity but could be constrained by habitat complexity (Briand and Cohen 1987) or ecosystem size (Post et al. 2000).

Analysis of 16 modestly well-resolved food webs with 25 to 172 nodes did not show scale-free structure (Dunne et al. 2002). There are, however, properties that set food webs apart from completely random networks (Montoya et al. 2006) and other properties (clustering patterns) that ecological communities share with large networks such as genetic responses, scientific citation networks, and internet links.

The prevalence of positive or negative interactions is another area of controversy in ecological community structure. For example, data suggest that the importance of facilitation or mutualism has been vastly underestimated (Bertness and Callaway 1994, Bertness and Hacker 1994). A survey of seven broadly different plant communities from desert to alpine (Tirado and Pugnaire 2005) indicated the importance of positive interactions and facilitation in six of the seven assemblages. For now, we do not have enough complete information on all interactions in large communities to establish patterns of relative importance of interactions across communities.

Community structure has been investigated by analyzing patterns of competition or of food webs across communities (e.g., Pimm 2002). The question remains whether such a conceptual approach has bearing on real communities if it does not consider mutualism, facilitation, amensalism, or commensalism. Food webs are the central avenues of energy flux through communities, and patterns of trophic interactions could thus provide the dominant structure in ecological communities. However, the dominance of food webs in structuring communities has not been tested against alternative models that also include all other interactions that can be found in communities. For example, many terrestrial food webs would collapse or change drastically if mycorrhizae and pollinators were removed.

Controversy over the stability of communities as a function of the number of species in the community (McCann 2000) has occurred since the beginnings of modern ecology (e.g., MacArthur 1955, Elton 1958, May 1972, May 1973, Frank and McNaughton 1991). Diversity and stability are discussed in greater detail in Chapter 3 but briefly considered here. Diversity and stability linkages are not yet highly predictable in a practical sense. Part of the reason for this unpredictability is that neither diversity nor stability has simple definitions (e.g., the term *disturbance* can mean resilience, resistance, persistence; *diversity* could be richness or evenness).

A practical example of the inability to link diversity to stability comes from the literature on stability as exemplified by the resistance to invasion. The ability of nonnative species to establish in communities could be lower in a more diverse community (e.g.,

Stachowicz et al. 1999, Zavaleta and Hulvey 2004). Levine and D'Antonio (1999) found limited empirical support for a positive relationship between diversity and resistance to invasion in field and smaller scale studies and weaknesses in the theory that leads to these predictions.

The relationships between stability and complexity of communities are difficult to resolve in part because of the paucity of data on the strength of interactions (Paine 1992, Montoya et al. 2006). Although mathematical methods to determine stability of interaction matrices exist (e.g., loop analyses, Bodini 1998), these methods may miss stability that is based on tightly interacting clusters of species in a network of loosely connected clusters.

The idea of keystone species has been well received in ecology since Paine (1966) introduced the concept (the actual term was introduced in a later publication). A keystone species has been defined as one whose effect is disproportionately large relative to its abundance (Power et al. 1996). If there is a distribution of species strengths, some species must be stronger interactors. A general predictive theory to predict which species in a community will be a strong interactor has remained elusive. Predicting *a priori* which species will be a keystone is difficult without specific knowledge of the natural history of the community. To make things more difficult, species that are not strongly connected to the community can have strong effects when they are removed (Berlow 1999).

Island Biogeography

Island biogeography is the theory of how size of a habitat and distance from a source of colonists can control the diversity of an island. Island biogeography was initially applied to islands in the ocean but can now refer to any patchy habitat in a hostile matrix. General patterns that can be explained from the model include decreased diversity with smaller island size or more distance from colonization sources. Island biogeography can be considered the simplest case of landscape ecology. The idea that island biogeography is a theory has been discussed in detail (Pickett et al. 1994).

Laws and Theories Most Applicable Population heterogeneity. Scaling. Heterogeneity increases diversity. Diversity positively correlated with area.

Strong Patterns

- Smaller islands have fewer species, and those species tend to be less competitive than species from larger islands or continents. Thus, new invaders often can extirpate existing species.
- From the theory of metapopulations, small populations are more likely to go extinct.

Some Predictions Island biogeography is an area of macroecology featuring models derived from population-based mechanisms (extinction and colonization) that has been successfully used to predict spatial patterns of diversity. Initial concepts behind the theory were based on colonization of oceanic islands (MacArthur and Wilson 1967). The theory is based on reasonable and sound assumptions; fewer colonists find islands when they are farther away from a colonization source, smaller islands have fewer subhabitats and support fewer species, and the probability of extinction is greater in smaller areas. Early experiments confirmed that equilibrium is reached on islands where immigration and extinction rates balance (Simberloff and Wilson 1969).

Island biogeography has been extended to other insular habitats that are not necessarily surrounded by water. Places where species-area relationships have been demonstrated include diatoms on slides placed in streams (Patrick 1967), fish in small lakes (Tonn and Magnuson 1982), invertebrates on stones in rivers (Douglas and Lake 1994), and small mammals on mountaintops (Brown 1995).

Species-area relationships combined with the insular nature of islands lead to predictions about the rate new species are accumulated per unit area increase (Rosenzweig 1995). For example, continental areas will have a steeper rate of increase in species accumulation curves as more area is sampled than islands. This prediction of island biogeography has been well supported with data.

Some Difficulties with Prediction A difficulty with island biogeography is predicting extinction rates and rates of successful establishment. No mechanism is available to predict extinction rates in most environments, other than they increase as population size decreases. Thus, many aspects of island biogeography are derived empirically, rather than mechanistically.

Pickett et al. (1994) list some potential questions with the theory in regard to generalizing its predictions. First, what is an effective island? How insular does a habitat need to be before it is considered an island? An island in the Pacific Ocean could be an effective island for most plants and animals but have the same complement of cosmopolitan protozoa found on all continents. Second, how does this theory underlie other biogeographic determinants of species abundance? For example, islands of the same size, age, and distance from the mainland can vary in species diversity if one is in the Arctic and the other in the tropics. The tropical island might have a greater diversity of colonists because broad biogeographical patterns dictate that the source of colonists is much more diverse in the tropics. The number of terrestrial species on Antarctica is much less than Australia, even though Antarctica is several times larger than Australia and Antarctica had many species when it had more temperate conditions (50 million years ago). Harshness (the law that organisms require liquid water) overrides the species-area effect. Third, how do characteristics on individual islands differentially affect species extinction and invasion? Fourth, can the theory be generalized to situations with gradients of hospitable habitats around the island? Fifth, how do spatial scales and landscape characteristics alter predictions of the theory?

The lack of prediction capability of island biogeography theory could explain variance in the relationship. Species-area relationships often only explain about 50% of the variance in species richness and thus may not be much use for predictive management except in the rough sense that larger areas will have more species (Simberloff 2004).

Some invasive species colonize and drive existing species off islands. Invasive species have a greater probability of success colonizing islands than continents. Species from larger islands or continents have evolved to successfully interact with many other species, particularly in competitive or exploitative situations. Species on islands evolve adaptations for survival on islands, perhaps more toward intraspecific than interspecific competition. Some of the characteristics evolved on islands (e.g., flightless birds) make species more susceptible to predation. Species introduced to islands have repeatedly caused dramatic decreases in biodiversity including but not limited to humans on Australia and New Zealand causing extinctions of most large animals, the brown tree snake in Guam extirpating many species of birds, and introduced birds and plants in Hawaii displacing most of the native fauna and flora. Humans are a force that makes prediction particularly difficult, because they often decrease diversity by removing species on islands or introducing species that outcompete many native species. However, humans increase the rate of colonization by new nonnative species (i.e., introduce plants and animals) and can actually increase the overall diversity of an island.

Species-area relationships can be combined with models of trophic interactions with the goal of explaining numbers and types of trophic links across landscapes (Brose et al. 2004). However, these models are based on power scaling of trophic links among species, which remain controversial. Trophic rank could also influence steepness of the relationship with a greater increase in diversity of predators as area increases than found for primary consumers or primary producers (Holt et al. 1999), but data for this relationship are modest with some support from insectivorous lizards and zooplanktivorous fishes (Ryberg and Chase 2007). Thus, associations between trophic links and island biogeography do not lead to predictions at the level of a theory but may provide patterns that can apply to specific situations.

ECOSYSTEMS

Abiotic factors, which were prominent in physiological ecology, attain increased importance again in theories of ecosystems. I also use this section to illustrate that some theories can be constructed with specific constraints (unidirectional stream flow leads to the theories

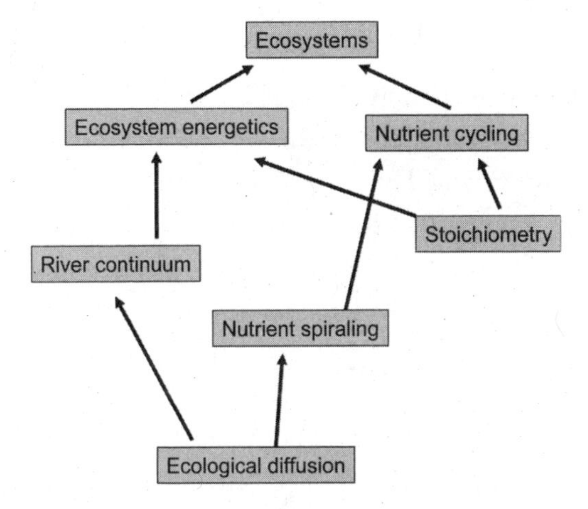

Figure 12. Ecosystem ecology and related theories as discussed in this chapter. Stoichiometry is useful for tying nutrient cycling to ecosystem energetics. Two more limited theories presented here are constrained by unidirectional flow in streams (nutrient spiraling and the river continuum).

of river continuum and nutrient spiraling) to illustrate how habitat-specific theories can increase ecological predictive ability. Defining the appropriate boundaries of ecosystems can be difficult (Post et al. 2006), so it will be left that boundaries are defined by investigators choosing the relevant sizes of systems of study. Ecosystem energetics, nutrient cycling, stoichiometry, and ecological diffusion combine to provide part of the structure of ecosystem ecology (Fig. 12). Nutrient spiraling and the river continuum theory are discussed as examples of more restricted ecosystem theories that I am comfortable with to illustrate how specific constraints can be combined to make narrower but more predictive theories.

Control of Ecosystem Energetics

Since Lindeman (1942) published his seminal article viewing the ecosystem from an energetic viewpoint, ecosystem theory has been intertwined with energetic considerations. Lindeman's paper quantified the standing stocks of energy and the efficiency of transfer of this energy through the food web. The concepts were then picked up and developed into a full theory by several scientists, with crystallization and widespread dissemination of the ideas in E. P. Odum's *Fundamentals of Ecology* first published in 1953. The approach was to use simple budgeting principles and physical laws to build a model of the ecosystem as characterized by flow pathways, efficiencies, and storage of energy. The concept of ecosystem energetics was moved forward by quantifying the flux rate of energy among compartments in the food web (Wiegert and Peterson 1983). What is the limiting factor or factors (primary constraint) for the ecosystem process of interest? If this question can be answered, then substantial predictive ability is obtained via ecosystem theories.

Laws and Theories Most Applicable Laws from physics and chemistry (particularly thermodynamics). System openness. Recycling rates. Energy requirement. Maximum metabolic rates. Limits to growth. Temperature optimum. Biotic/abiotic interaction. Evolution affects ecology. Scaling.

Strong Patterns

- Ecosystems are structured by the fact that the primary energy flux is dominated by primary producers or heterotrophic decomposition of allochthonous organic materials.

Some Predictions Ecosystem energy budgets are quantified fairly easily because the budgets (input, output, and storage) must balance; inputs, outputs, and storage for each compartment must balance as well. The requirement for balance allows researchers to characterize energy budgets for whole ecosystems and make cross-ecosystem comparisons of relative system productivity and efficiency. Identification of the limiting factor or factors as a constraint facilitates prediction of ecosystem rates.

It is possible to have more biomass at a higher trophic level than a lower level (an inverted biomass pyramid, Odum 1959), but total energy flux into biomass (turnover) must decrease with each trophic level. Carbon flux is generally considered equivalent of energy transfer, and this is mostly true, particularly for consumers.

In many ecosystems, it is possible to predict initial energy flux by knowing constraining abiotic factors. For example, maximum net ecosystem production can be predicted in many lakes if the influx of nutrients and amount of light attenuation affiliated with nonliving materials is known. By extension, fish production can be significantly correlated with P loading rates to a lake (Fig. 13). This relationship could be viewed as P constraining maximum potential productivity, but some systems have lower productivity based on influence of other limiting factors (e.g., N, turbidity that attenuates light, high rates of fish harvest, and existence of trophic cascades that alter energy transfer from phytoplankton to zooplankton). Thus, for any upper value of fish yield in Figure 13, there are other values up to 10 times lower. An additional constraint is that fish production will decrease or fall to zero in extremely eutrophic systems where anoxic conditions lead to complete loss of fishes. The relationship between P and fish production is a good example of predictions that can be made

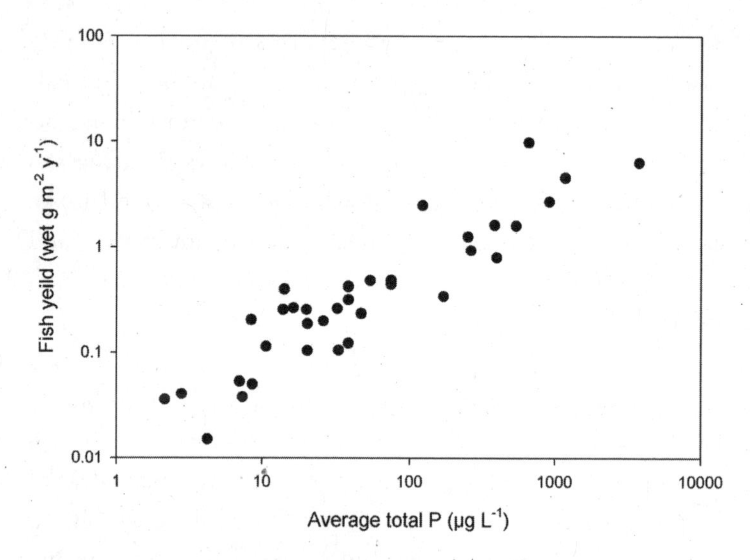

Figure 13. Relationship between average total P concentration and fish yield for lakes. Phosphorus concentration is calculated as a function of influx of P to lakes and individual lake residence time assuming that sedimentation and recycling from sediments are in an approximate balance. Plot is from Lee and Jones (1991).

by the theory of ecosystem energetics. First, primary production rate (energy capture rate) is limited by nutrients. Second, the rate of primary production limits fish production. Thus, we expect a positive relationship between increased total P and fish production.

In terrestrial ecosystems, precipitation can be used to predict more than one-third of the variance in primary production (Fig. 14A). Additional information (e.g., temperature, timing of precipitation, evapotranspiration) can be used to establish better predictive estimates (Fig. 14B). Relationships between precipitation

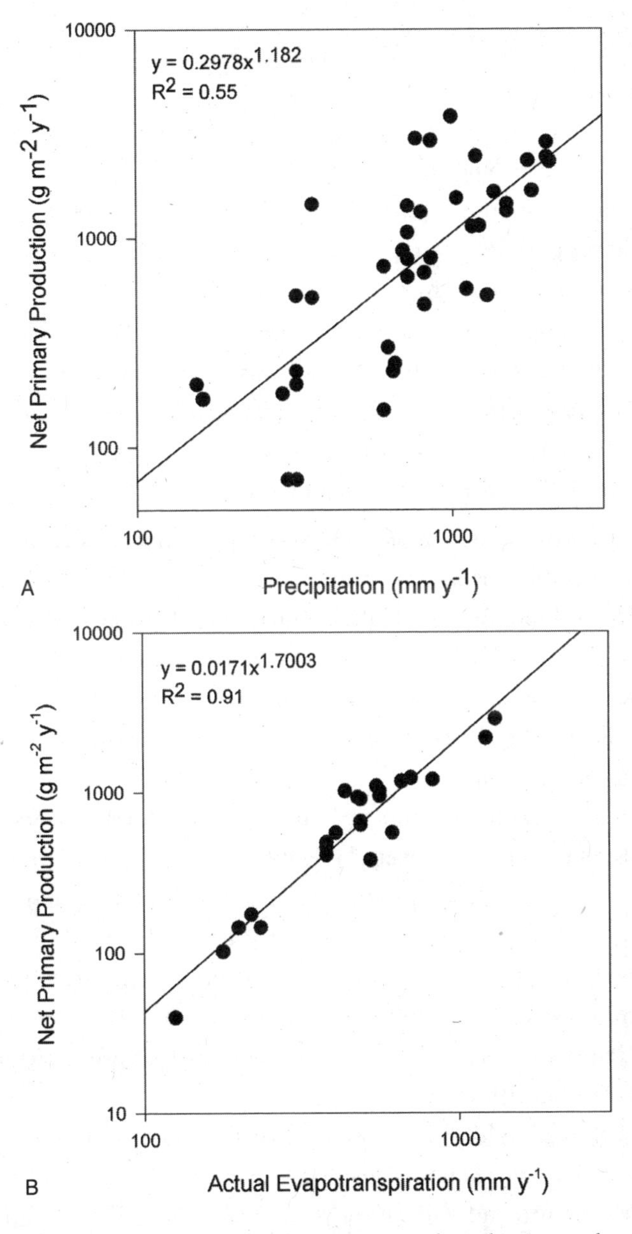

Figure 14. Relationship between annual production and precipitation for a number of tropical ecosystems (A, data from Murphy 1975) and net primary productivity as a function of actual evapotranspiration for terrestrial communities ranging from arctic to desert (B, data from Rosenzweig 1968).

and production are not purely empirical predictions; they are based on physiological properties of plants.

One question this type of plot (e.g., Figs. 13 and 14) raises is, what proportion of variance should be explained in order to say that a theory is useful? This is a separate question from the statistical significance of the trend. There are good reasons that precipitation does not exactly relate to production including temporal variance (if it rains when organisms are inactive they do not grow) and variance in nutrient limitation. Ecologists seem happy if half the variance can be explained.

Some Difficulties with Prediction Accounting for energy based on carbon has problems because some sources of carbon (e.g., tannins, lignins, cellulose) are so recalcitrant that the efficiency with which they can be used by organisms varies widely. This efficiency effect is not as great for predators as herbivores because animal prey is less likely than plants to contain a substantial amount of recalcitrant carbon compounds.

Beginning ecology students learn the 10% rule (90% of energy is lost at each trophic level). However, some organisms are substantially more efficient with the energy they obtain, and others are less efficient. For example, net production of herbivorous insects relative to net production of plants in the same area ranges from less than 1% to 19% with a mean of 3.5%. Production of predatory insects averages about 20% of ingested prey (Wiegert and Peterson 1983).

In the example of the relationship between P loading and fish production, we cannot predict the exact rate of fish production from first principles given the P loading. The efficiency of energy capture per unit P loading, the degree of omnivory in the food web, and the energy transfer efficiency per trophic level in the food web are all variable enough that the precise amount of energy transfer to fish biomass cannot be predicted.

Length of food webs was thought by some to be limited by primary production (Hutchinson 1959), but analyses of numerous food webs has indicated that this is not correct (Briand and Cohen 1987). Efficiency of energy transfer and basal production rates does not necessarily constrain the number of trophic levels. One reason for the mismatch between production and food chain length could be that more variable populations lead to shorter food chain length (Pimm 1991).

Nutrient Cycling

A theory of nutrient cycling is not difficult to build because so much of the basic study of nutrient cycling by ecosystem ecologists is based on well-established physical and chemical laws. Much of the work on nutrient cycling has centered on budgeting fluxes and standing stocks of nutrients based on the law of conservation of mass. A main goal of nutrient cycling theory is predicting flux rates of nutrients. The theory of nutrient cycling has application for understanding how ecosystems work and also how human nutrient pollution and interference with inputs will influence natural ecosystems. Finally, nutrient flux through ecosystems is driven in large part by microbes (e.g., ruminants, mycorrhizae, microbial decomposers, and microbial producers in aquatic ecosystems), so given the spatial and temporal scales on which humans observe nutrient dynamics, the ecosystem behaves as a large-number system.

Laws and Theories Most Applicable Laws from physics and chemistry (e.g., nutrient transport, gas flux, solubility in water as related to weathering and precipitation of chemical forms, chemical interactions related to diagenesis). System openness. Recycling rates. Energy requirement. Nutrient cycling requirement. Maximum metabolic rates. Water requirement. Temperature optimum. Biotic/abiotic interaction. Evolution affects ecology. Scaling.

Strong Patterns

- From the law of diversification and evolution, the theory predicts that it is unlikely for molecules with potential energy to build up to large stocks in the environment, particularly those containing elements essential for biochemical processes central to life (e.g., N, P, S, C). This strong pattern has been stated previously "Over geological time life forms evolve to utilize a wide range of novel resources, often biologically produced, such as oxygen, lignin, cellulose and petroleum hydrocarbons. If a resource appears in the environment, then it seems organisms will evolve to use it" (Wilkinson 1999, p. 535).
- Heterotrophs, with the exception of many detritivores or grazers that rely on carbon-rich food, tend to mineralize organic materials releasing inorganic nutrients.
- Autotrophs tend to incorporate inorganic nutrients and convert them to organic form.
- The vast majority of food webs depend on energy initially derived by photosynthesis either inside or outside the system. A very small minority depend on chemoautotrophic organisms.

Some Predictions The strong pattern, that nutrient cycles are complete, leads to prediction of transformations resulting in complete nutrient cycles because it is unlikely for chemicals with potential energy to build up to large stocks in the environment, particularly those essential for biochemical processes central to life. Most theoretically possible nutrient transformations that would be biologically advantageous and involve moderately to highly abundant compounds have been found in nature. A potential exception is using phosphate as an electron acceptor for respiration, but this would yield toxic phosphine, PH_3, a highly flammable gas. There are some recent examples of fluxes predicted by the theory of nutrient cycling related to the concept that any time there is potential energy in the environment in chemical forms, organisms will evolve to use it. These examples include the energy yielding pathway of oxidizing ammonium with nitrite (Dalsgaard and Thamdrup 2002) and disproportionation of thiosulfate (Jørgensen 1990).

Nutrient cycling is driven by the different requirements of autotrophs, detritivores, herbivores, and other consumers. The ideas of complete nutrient cycles and stoichiometric constraints lead to the observation that consumers tend to mineralize organic materials releasing inorganic nutrients. This mineralization occurs because the composition of organisms is constrained within ranges, and heterotrophs require organic carbon not only for growth but also for energy. Heterotrophs have a greater carbon requirement than autotrophs because they need carbon for energy as well as to build cells. Autotrophs and detritivores relying on carbon-rich food sources incorporate inorganic nutrients and convert them to organic form. Thus, carbon flux through food webs can ultimately be controlled by rates of input of nutrients to an area from outside, but mineralization within the area can substantially increase rates of carbon flux through the food web.

As stated under strong predictions, the vast majority of food webs depend on energy derived by photosynthesis, and this is a strong pattern. Photosynthesis driving food webs is a predictive statement that is for most practical purposes true but may have some interesting exceptions. Even communities, such as those based on sulfur oxidation (e.g., deep-sea vents, some cave ecosystems, Hutchens et al. 2004), are ultimately dependent on dissolved oxygen derived from photosynthesis (i.e., on the potential energy gradient created when oxygen-containing ocean water that has been oxidized by photosynthetic organisms comes in contact with reduced compounds dissolved from Earth's crust). In any case, communities that depend on chemoautotrophy are very rare relative to those driven by autotrophy. The only exceptions to ultimate direct or indirect reliance on carbon fixed or oxygen released by photosynthesis could be deep subsurface microbial communities based on methanogenesis driven by hydrogen and carbon dioxide released by geological processes (Chapelle et al. 2002).

Basic chemical laws (e.g., solubility in water as related to weathering and precipitation of chemical forms, chemical interactions related to diagenesis) and physical laws (e.g., nutrient transport, gas flux) are also important in the theory of nutrient cycling. Application of these laws allows prediction of chemical reactions with biotic implications. For example, at high pH, carbonate and phosphate will precipitate, potentially leading to P limitation. A second example is the anoxic hypolimnion of some lakes; ferric phosphate will dissociate and allow phosphate to be mixed throughout the lake when stratification breaks. This is why it is difficult to reverse eutrophication once a lake has reached a point where the hypolimnion becomes anoxic (Dodds 2002). A third example is that acid precipitation can have only modest effects while the soil maintains its buffering capacity. Once that capacity is exceeded, rapid decreases in pH of soils and downstream waters occur, and a cascade of biological effects follows.

A powerful prediction of nutrient cycling theory is the order that electron acceptors will be used in anaerobic respiration. This prediction is made based on redox potential energy and yield of chemical reactions involving oxidation of organic carbon. Thus, aerobic respiration will predominate until dissolved oxygen is depleted, denitrification (nitrate used to oxidize organic carbon) will follow, then manganese (Mn^{4+}, Mn^{3+}), ferric iron (Fe^{3+}), and finally oxidation using sulfate (SO_4^{2-}). Only after sulfate is depleted will methanogenesis (combination of carbon dioxide with hydrogen to yield energy and methane) and related reactions occur at significant rates (Westermann 1993).

Feedbacks between biota and abiotic habitat are another important aspect of the theory of nutrient cycling. Such feedbacks include global climate regulation and microbial involvement in weathering materials. Global climate regulation does not require evolution at universal levels but can be a consequence of individual species evolving to take advantage of chemicals built up in the environment that have substantial potential energy (Dagg 2002). Mineralization of nutrients mentioned at the beginning of this section is another of these feedbacks. Feedbacks, nonlinear responses to controlling factors, and variable storage rates all increase predictive difficulties.

Some Difficulties with Prediction It is difficult to predict the absolute flux rates of many nutrient cycles. For example, nitrate uptake in streams across large biogeographic areas varies over many orders of magnitude. Only about half the variance in absolute rates can be explained even though many factors (e.g., temperature, metabolism, light, nutrient limitation) have been measured (O'Brien et al. 2007, Mulholland et al. 2008). The variance could be explained by complex interactions among these factors or factors not yet measured. These estimates of nutrient cycling and retention are important to understanding how humans are altering the global N cycle because instream N retention could be an important control on the rate of movement of N from land to oceans (Bernot and Dodds 2005) and, therefore, the translocation of fertilizer from land to aquatic environments.

Similar problems occur in determining rates of carbon storage across global ecosystems and how they will change in the future. Carbon storage needs to be quantified to explain the relationship between carbon input to the atmosphere and atmospheric carbon dioxide concentrations. Currently, the global carbon budget is not completely balanced (e.g., Plattner et al. 2002). Conservation of matter requires a balance; the difficulty of measuring rates of ecological processes and scaling to global rates has hindered complete description of the global carbon cycle. Interactions among nutrient cycles on a changing planet subject to rapidly increasing anthropogenic influences also increase the difficulty of prediction.

Nutrient cycles do not occur in isolation; some interactions are, at present, too difficult to predict. There is still difficulty predicting why N should ever limit primary production because autotrophs (cyanobacteria) and other heterotrophic bacteria (e.g., *Rhizobium, Azotobacter*) are able to fix atmospheric N. This limitation may be related to molybdenum dynamics (Howarth and Cole 1985) or the fact that N fixation is a very energy-demanding process and some biogeochemical aspects favor N movements differently than P (Vitousek and Howarth 1991). The relative importance of these various controls has not yet been unequivocally demonstrated for global-scale N cycling.

Forecasting global nutrient cycles is difficult because of feedbacks and nonlinearity associated with directional global change. Again, the global carbon cycle is difficult to predict. Global warming could melt high latitude permafrost and lead to greater rates of respiration and methanogenesis. A substantial portion of carbon stored in soils on Earth is frozen in northern North American and Asia. If that carbon is released as radiatively forcing gasses, greenhouse warming could accelerate leading to even more rapid release of greenhouse gasses.

Specific controls over many local ecosystem nutrient cycling rates are difficult to predict with a high degree of accuracy. Directional global change, unprecedented levels of nutrient addition, interactions with organisms, deforestation, and establishment of introduced species (Vitousek et al. 1997) all make such prediction more difficult.

Nutrient Spiraling and Solute Transport

All laws that are most applicable to the general theory of nutrient cycling apply to this law. This theory provides an excellent example of application of a fairly straightforward set of predictions following specific constraints that occur in a single type of habitat; unidirectional transport of materials through an ecosystem leads to specific predictions related to this one overriding contingency. This theory has primarily been applied to streams, but any ecosystem with directional transport (e.g., groundwater flow, water flow through a soil column) can be analyzed with similar mathematics and approaches. This theory can also be considered an extension of the theory of ecological diffusion discussed at the beginning of this chapter.

Laws and Theories Most Applicable All laws from the theory of nutrient cycling under the added constraint of unidirectional flow.

Some Predictions Webster (1975) made the simple observation that in streams, unidirectional flow of water moves materials in dissolved phase downstream more rapidly than when they are in particulate phase. The fact of advective transport leads to the idea that nutrients spiral rather than repeatedly cycle in one location, and leads to mathematical equations that can describe spiraling (Newbold 1981). The primary conceptual advance made in this theory over the theory of nutrient cycling is that it considers the consequences of movement of nutrients through an open ecosystem with unidirectional transport. Most previous approaches to nutrient cycling consider the ecosystem as a closed and well-mixed reactor with inputs and outputs.

The theory predicts an exponential decline of the probability that an individual molecule of a nutrient will move downstream. The prediction is verified frequently in streams where tracer releases are used to characterize nutrient dynamics (Fig. 15). These data illustrate the dynamic nature of nutrients in directionally flowing shallow benthic systems because the average molecule of nutrient is removed from the water column in a few hundred meters, which often corresponds to less than an hour of downstream travel time.

The theory of nutrient spiraling in streams is driven by hydrologic properties of streams (physical laws dictate that water flows downhill through streams) and by the requirement for organisms to transform nutrients to survive and grow. Further

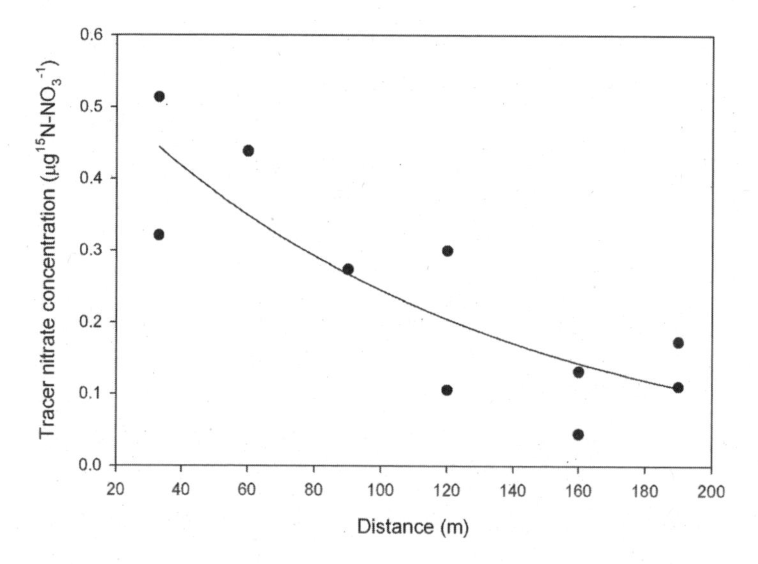

Figure 15. Measured decline in a N isotope tracer over distance in a small prairie stream corrected for dilution fitted with the expected functional relationship predicted by spiraling theory, an exponential decline (unpublished data).

advances in this theory have come through hydrologic studies that characterize the movement of inert solutes through streams (Stream Solute Workshop 1990). The idea that some parcels of water in a stream are retained more stringently than others lends complexity to the model (e.g., groundwaters near the stream or large pools in the stream with slow water replacement time retain materials).

Equations relate nutrient spiral lengths to physical properties (Stream Solute Workshop 1990) including

$$S_w = uzC/U \quad (16)$$

where S_w is the average distance a dissolved nutrient stays in the water column, u is water velocity, z is average depth, C is concentration of the nutrient, and U is uptake (either biotic or abiotic) of the nutrient per unit area of stream bottom. Some straightforward predictions from this theory as exemplified by equation 16 include that uptake length should increase with greater water velocity, greater water depth, greater instream nutrient concentration, and less active producer or heterotrophic microbial activity on the stream bottom. In streams dominated by algal production, uptake length can be tied to light because photosynthesis leads to greater nutrient demand. Spiraling theory can be used to predict movement of nutrients through networks of streams (Ensign and Doyle 2006). Spiraling theory has led to specific predictions of nutrient retention rates (Peterson et al. 2001, Mulholland et al. 2008) that can be directly tied to downstream water quality. In practical terms, this approach can be used to characterize the downstream influence of a plume of nutrients entering a stream (e.g., sewage effluent). The theory has also been combined with stoichiometric theory to predict that downstream transport of nutrients will be less when carbon input is high because carbon input leads to more retention (Webster et al. 2000).

Some Difficulties with Prediction The theory cannot currently predict temporal variance in nutrient transport, long-term geomorphologic depositional patterns of nutrients, and effects of biota, such as top-down control. Nutrient retention can be saturated, so uptake length becomes essentially infinite, but the exact form of this saturation for all nutrients is not known. Progress has been made for saturation effects on nitrate transport where it is likely that an exponential relationship between nutrient concentration and uptake occurs with the value of the constant in the exponent being less than one (O'Brien et al. 2007).

Streams are notoriously nonequilibrium systems, and floods and droughts commonly cause drastic changes in the biota responsible for nutrient uptake. Floods and droughts can also load nutrients into the system at variable rates. Predicting the exact uptake length in a specific stream is difficult without additional information, although nutrient concentrations in the water provide a substantial amount of predictive ability (Dodds et al. 2002, O'Brien et al. 2007, Mulholland et al. 2008).

River Continuum

The river continuum concept (Vannote et al. 1980) has had tremendous predictive effect in stream ecology, and it is treated here as a theory. This theory is certainly less broad than others presented in this chapter because it applies to a single habitat type, but it does have some important illustrative value; the ability to make sweeping ecological predictions based on linkages of ecosystems and directional flow that characterize lotic networks provides a positive example of the predictive ability in some branches of ecological science.

Laws and Theories Most Applicable All the laws from the theories of nutrient cycling and the theory of community structure hold. The particular constraint that makes this a powerful theory is that water flows downstream in rivers and streams (directionality of *System openness*) leading to upstream to downstream connections.

Strong Patterns

- Vannote et al. (1980) made the observation that linkages between terrestrial habitats and drainage networks and subsidies from upstream are vital controls on river and stream ecosystems and communities.

- Primary production is low in shaded headwaters, greater in open midsized streams and small rivers, and less again in larger turbid rivers (Bott et al. 1985).

Some Predictions The original paper (Vannote et al. 1980) referred to the dynamic equilibrium of the abiotic template of rivers and streams as the driver of a dynamic equilibrium of evolved life history characteristics of organisms. The authors directed the original paper toward predicting characteristics of organisms found at different locations within a large river network, and this was extended to the management implications of disturbing long-term equilibrium to which species have evolved in response (Vannote 1981). These aspects of the original paper have mostly been ignored, and scientists have primarily focused on the abiotic drivers and their influence in a downstream direction (Fig. 16).

Given the type of riparian vegetation in small streams (closed canopy, open canopy, seasonal or not) the amount and timing of light and organic detritus entering the stream can be predicted. If light is high (no canopy, low turbidity), primary production by attached algae or macrophytes (autochthonous production) will be relatively important in the food web. If light is low, the system will depend on organic carbon inputs from terrestrial vegetation and transported from above (allochthonous materials).

The food source dictates the type of food available and the functional community composition. In areas dominated by leaf inputs, invertebrates that can process dead leaves and the microbes associated with them are abundant and productive. In areas where attached algae dominate, invertebrates that can scrape

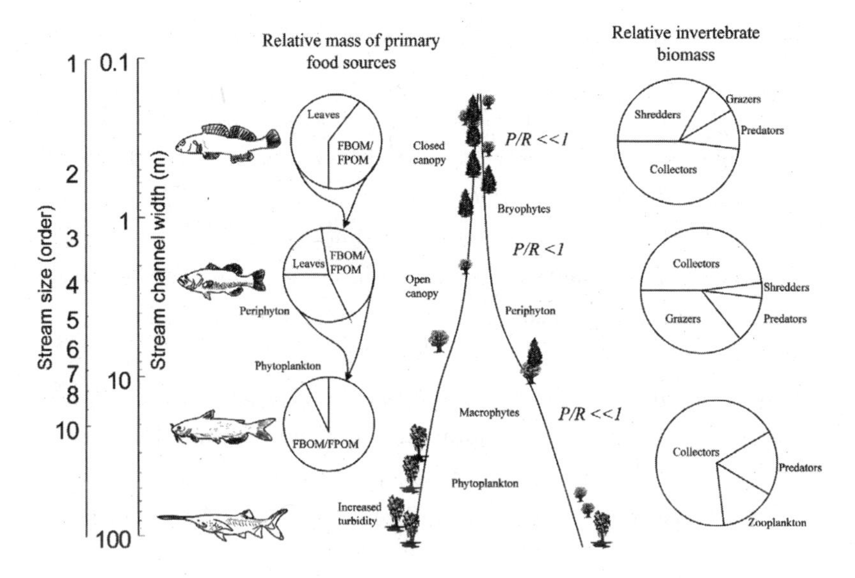

Figure 16. A conceptual diagram of the River Continuum Concept. The original diagram was published by Vanotte et al. (1981), this version is modified from Dodds (2002). Organisms are not drawn to scale. As stream size increases, the influence of the riparian canopy decreases and the influence of movement of particulate materials from upstream increases. In small forested streams the light is low and leaf input high leading to low rates of primary production (P) relative to respiration (R). Organic food sources vary with distance from headwater streams. The abiotic template allows for numerous predictions to be made ranging from ecosystem properties to life history characteristics of dominant species.

the algae from solid surfaces are common. In large rivers with little light, the community relies on dissolved and particulate carbon inputs from upstream, and predators adapted to large rivers (low light, high current, and muddy bottoms) are common. An example is catfish that depend on barbels to forage based on odor and well-developed lateral line systems to sense electric currents generated by prey.

This theory has been criticized because it does not predict what occurs in grassland, desert, or tundra river systems, but this is unfair criticism. The authors of the river continuum paper allowed for adjustments by biome (Vannote et al. 1980) but centered their arguments on deciduous forested streams. Although the authors humbly called it the river continuum concept, the level of prediction and its basis on sound laws merits consideration as a theory. It forms the theoretical basis and context for more recent synthetic efforts including the river discontinuum hypothesis (exploring the effects of human made impoundments on ecosystems, Ward and Stanford 1983) and the river ecosystem synthesis (Thorp et al. 2006).

Some Difficulties with Prediction As with any popular idea (the ISI web of science has the primary article cited more than 2,000 times by 2008), a number of publications have pointed out problems with the concept. These potential problems are mostly related to spatial patterning effects (although the original publication implied these effects). For example, a small stream can empty into a large river or ocean. Effects of stochastic events, such as floods and droughts, vary from upstream to downstream, but this has not been addressed. Another aspect of spatial patterning that may be important that is not directly accounted for is geomorphological patches that serve as functional units (for example, when rivers flow through heterogeneous geological areas). An example is the habitat created where tributaries intersect main stem rivers. These tributaries create patches that are not consistent with an orderly downstream progression (Montgomery 1999), and this spatial heterogeneity could have ecological significance (Benda et al. 2004). Patches could structure stream communities and ecosystems more strongly than the downstream position along a river continuum (Thorp et al. 2006).

Lakes form discontinuities in the continuum, and reservoirs form particularly unnatural and unpredictable discontinuities (Ward and Stanford 1983). Reservoirs are mostly problematic with regard to prediction because their water levels and releases vary based on human-driven needs, such as irrigation water, power generation, maintaining downstream flow for watercraft, and avoiding floods.

Because there is generally at least some attached algae, the presence of invertebrates that depend on algae is not precluded, even in highly shaded streams (Greenwood and Rosemond 2005). The taxonomic identity of individuals in functional groups (grazers, shredders, collectors) cannot be predicted without specific information (e.g., life histories or evolutionary zoography) on the stream being studied. The omnivorous nature of some stream organisms and their ability to rely on several types of food interferes with prediction. For example, fishes commonly assumed to be herbivorous can eat detritus and consume and assimilate animals (Evans-White et al. 2003). With omnivory, energy flux pathways in the food web are not fixed and alteration of food sources will not change the species composition of consumers. As with all theories, solid understanding of where prediction is limited allows appropriate application of the theory.

Probably the largest lack of predictability, as with all of ecology, is how organisms will evolve in response to the abiotic patterns they are subjected too. Turbid waters will provide strong selective force for nonvisual sensory adaptations, a straightforward prediction how nonvisual sensory systems will evolve can not. Take for example the catfishes and knife fishes in the Amazon (Stoddard and Markham 2008). Electrical currents are used to sense prey. Other species emit electrical currents as sort of radar to locate prey, but this makes them more "visible" to their predators, so the electrical fields are generated in such a way that they are not as detectable at a distance. The extreme is the electric eels that shock their prey to immobilize them. Different electric frequencies are emitted to evade detection by predators and these signals are used by some species to communicate. This degree of adaptation to lack of light is fascinating, but the specific adaptations are not predictable based on the River Continuum Theory.

CONCLUSION

This chapter discusses some potential examples of theories in ecology based on my proposed laws and strong observations. For each of these theories, I attempted to define boundaries with regard to the ability to use them to predict outcomes; but given the defined constraints, each theory allows for an acceptable degree of exception. Of course, my view of what constitutes a reasonable exception rate is subjective, and readers must decide if they agree or disagree with my specific view on theories in ecology. New general theories in ecology will be added to this list and probably some obvious theories already in the literature are not covered here.

In addition to these theories, a number of patterns in ecology are useful because they capture important features of some systems. The problem with these patterns is that they do not allow for wide-scale prediction. Often, these patterns were or are bandwagons; they generate substantial excitement and a flurry of publications when introduced (or more often rediscovered) but do not lead to prediction in the long run. Some of these patterns could ultimately have constraints and mechanisms so well described that they become theories. Other patterns will be discarded as useful conceptual models as the science moves forward.

CHAPTER 3

Patterns, Questions, and Predictions

Some questions have been driving the field of ecology since it began, and numerous concepts or "hot topics" in ecology have been explored in passing over the decades. Predictive ecological science related to popular areas of ecological research or areas of ecology that have been mostly ignored but seem important to me are considered here. Some questions presented below have properties leading to patterns that can be used for prediction because an ecologist should be aware of the potential for the patterns to hold in a particular habitat. But, it is not wise to assume the patterns described in this chapter hold everywhere, because these observed patterns are based on constraints that vary among many habitats. Therefore, these patterns are useful but not broadly predictive in the sense of laws and theories built from those laws. These patterns are presented as questions because of my uncertainty about the generality of the answers that the ecological observations provide. Questions cover some of the most active areas of research and are of interest to many researchers. Additional questions are posed related to the general way we approach ecological science and are considered in the first section of this chapter.

GENERAL QUESTIONS

First, a series of general questions is approached. These questions are not constrained by hierarchical scale and are related to the ability to perform predictive ecology. Three of these (the utility of empirical and mechanistic models, the utility of neutral models, and the role of natural history) address how we actually do ecological research in a predictive sense. The last two general issues (questions of scale and applicability of fractals in ecology) are related to how scientists can deal with the tremendous range of spatial and temporal scales across which organisms operate.

What Is the Relative Utility of Empirical and Mechanistic Models?

This section is inspired by the provocative book by R. H. Peters (1991) that can be crudely characterized as arguing for empirical approaches to ecology and against mechanistic ecological models that were viewed as useless for prediction. This approach was discussed in the Introduction, but it is worth revisiting now that a number of laws and theories have been considered.

Empirical models (based on statistical relationships using measured data) have great utility in predictive ecology, but they may not predict by extrapolation outside the range of previously experienced conditions. Empirical models also have a substantial amount of variance. Many ecological papers, including several of mine, have presented empirical relationships that are highly significant, but regression can explain less than one-third of the variance. Mechanistic models are more difficult to come by but may account for deviations from prior conditions. The balance between empirical and mechanistic models is discussed with respect to a few successful predictive models to illustrate the balance between mechanistic and empirical approaches and how both are generally present in modern ecological approaches.

The first model considered here links P loading to algal biomass in lakes. Phosphorus loading modeling of eutrophication has been a successful empirical approach because it is widely used to make management decisions, and results of management actions generally fall within prediction of the model. Although this model can be viewed as empirical and based on the observation of a strong correlation between total P and suspended chlorophyll across many lakes, it is also based on clearly understood mechanisms. The scientists creating and using the model did not happen on a chance correlation between P loading and chlorophyll concentration in lakes. They found the pattern because they hypothesized that increases in P stimulated phytoplankton growth. The model is highly predictive relative to many ecological models (often explaining more than 80% of the variance in observed mean chlorophyll concentration). Vollenweider (1976) solidified the factors controlling total P concentration in the mixed upper level (epilimnion) of stratified lakes using a mass-balance approach. This approach accounted for the volume of the epilimnion, the amount of P entering the lake per unit time, dilution of P by water flowing through the lake, and loss of P to sedimentation during the summer stratification period. The budgeting approach is driven by the law of conservation of matter from physics and the law of *System openness*. Mechanisms behind most of the variables used in the model are well understood. Chlorophyll yield per unit P, factors controlling depth of the epilimnion, settling rates, and dilution are well documented and understood. Yet, there is still variation related to other factors such as suspended sediments, trophic effects, and temperate versus tropical climates. Application of constraints to the model makes it better at predicting specific responses in lakes of more localized regions. For example, tropical lakes may have a longer growing season and respond differently to P loading.

The Michaelis-Menten model of enzyme kinetics (also known as the Henri-Michaelis-Menten model) is derived from first principles of enzyme kinetics. The model has received substantial empirical support from research on cultured organisms in the laboratory and works well with individual species. Entire microbial assemblages also can display uptake rates that follow the general form of Michaelis-Menten curves. The model underlies growth of organisms (Monod growth curves) because growth rates depend on enzymatic activity rates. Uptake is not the same as growth, so growth is indirectly related to enzyme kinetics, but also saturates as resources become more abundant. Saturation of growth can be combined with the fact that half saturation constants are not the same for all organisms (driven by the law that organisms must specialize) which leads to a theory predicting how resource ratios influence ecological communities (Tilman 1982). The Michaelis-Menten relationship is a model with a clear mechanistic basis for the specific equation that is used. The range of model parameters of individual species in an ecosystem cannot be predicted and must be determined empirically.

Photosynthesis-irradiance models are successful at characterizing relationships between light and photosynthetic rates and have both empirical and mechanistic features. These models have several features in common; they increase initially (with a slope of alpha) and they plateau at a maximum rate (i.e., display a hyperbolic tangent). Photoinhibition can occur in terrestrial plants when they are stressed or moved from a low-light to a high-light environment. Phytoplankton can be exposed to high light when shallow in the water column and experience photoinhibition. Several equations are used to describe photosynthesis-irradiance patterns for phytoplankton (e.g., Jassby and Platt 1976) that reasonably replicate relationships between light and photosynthetic rate. These equations are not, however, all based on first principles. The prediction of an initial linear increase in photosynthetic rate proportional to light is mechanistic; the process of photosynthesis has a specific quantum yield, and twice the light leads to twice the yield. Likewise, there is a maximum yield that leads to saturation. Just as enzymes have a maximum rate of activity, photosystems have a maximum rate at which they can work. However, the process of photoinhibition can set in once light intensity is great enough. There is no mathematical form that the process of photoinhibition should take because of the complexity of inhibitory processes at the molecular level and multiple potential causes for decreased photosynthetic rate caused by cellular damage from high irradiance. The empirical nature of a photosynthesis-irradiance model that fits photoinhibition at high light intensity does not preclude its use in ecology.

These three examples illustrate the potential applicability and interplay between empirical and mechanistic models. Empirical relationships can be important, because they start the quest for mechanisms. For example, the field of metabolic scaling illustrates strong empirical patterns that beg for mechanistic understanding.

Empirical models can also be important, because they allow for prediction without complete understanding of underlying mechanisms. The greatest strength of mechanistic models is evident when they represent processes not amenable to experimentation. For example, global warming and general circulation models fit into this category. We have only one Earth, so models of global climate are necessary because replicated experimentation cannot be done. Current increases in CO_2 are an experimental manipulation in a sense, and we are currently testing the ability of the mechanistic models to predict how much the earth will warm given the radiative forcing of increased greenhouse gasses. In general, a hybrid approach somewhere between mechanistic and empirical models seems necessary to provide prediction in ecology.

Are Null-model or Neutral-model Approaches Useful?

Null model and neutral-model approaches have become more common in ecological study. The idea that a minimal set of assumptions should be used as a null model to test against is a common philosophical approach to science (e.g., Alonso et al. 2006). This approach generally assumes that the principle of parsimony holds and the simplest explanation is usually best. The burden of proof is on the more complex hypothesis tested against the null hypothesis. The concept is a good one and often seems to lead to correct interpretations of nature. If so, some mechanism should be responsible for the difference between observation and the null model. Historic debates about community assembly were based on the idea that ecological patterns should be tested against a random distribution (null model). The neutral-model approach is to assume a random distribution of a specific property of organisms of interest in an ecosystem and ask if the observed pattern is significantly different.

In practice, establishing a neutral or null model is difficult because assumptions can hide biological mechanisms (Colwell and Winkler 1984). For example, competition tends to drive species away from an area or even to extinction, so current distributions may be influenced by competition that occurred in the past.

Commonly, null models are erected such that the null expectation is that the biological features of a system do not matter (e.g., all species are equivalent). However, other null models can also be constructed based on biological features. For example, a number of the laws presented in the Introduction are biologically based null models, such as the suggestion that all types of interactions can be found in communities.

The theory of island biogeography was an early-version neutral theory of spatially structured community assembly (May et al. 2007). This theory is neutral in a biological sense because it assumes nothing about the attributes of individual organisms or species. The theory bases its predictions on the idea of probability of colonization and extinction. Island biogeography has insularity as the primary contingency. A more advanced theory of island biogeography would take into account organism properties such as dispersal ability.

A neutral theory of biodiversity has been proposed by Hubbell (2001). His model relaxes one of the fundamental laws proposed in Chapter 1. As with island biogeography, he assumes ecological equivalence of organisms, which is in direct opposition to the idea that all organisms are unique. However, relaxing this assumption leads to distributions of biodiversity that often fit observed patterns and thus has received considerable support (Bell 2001).

The neutral theory of biodiversity generated a flurry of publications supporting and refuting it. A test of the theory's predictions on corals (Dornelas et al. 2006) resulted in significant differences from neutral-model predictions of community similarity and relative abundance of species. The lognormal distribution fit most data sets explored by McGill (2003) better than the distribution predicted by the neutral model of biodiversity. Experiments in grasslands yielded results that Fargione et al. (2003) did not think were consistent with neutral models, and rocky intertidal zone community dynamics were not predicted well with the neutral model (Wootton 2005). An analysis of 158 published data sets on community structure suggest that neutral processes were the only structuring processes in 8% of natural communities but that disregarding neutral dispersal would miss important community structuring processes in 37% of the cases (Cottenie 2005).

In contrast, Azaele et al. (2006) applied neutral-model approaches to data on trees in a Panamanian forest and found agreement of the model's predictions of relative species abundance, turnover, and extinction times. It is not too surprising that the model works well for this tree assemblage because Hubbell used data from the same forest to develop his original ideas. Neutral models describe bacterial communities in tree holes well (Woodcock et al. 2007). A study of forest fragments in the Amazon supported the predictions of species extinction but underestimated rate of change in species composition because the neutral model does not account for differential response of tree species to altered environments (Gilbert et al. 2006). The results are consistent with the review by Chave (2004), who found little empirical evidence for species trait equivalence but did find that relative fitness is approximately equal across species.

Latimer et al. (2005) found that neutral models explained biodiversity hot spots well for South African fynbo plant communities, but only if dispersal rates are substantially less than those observed in tropical forests. Inclusion of functional traits of plants (assumed to be equivalent in the neutral theory) considerably improved the ability to predict relative abundance of species. In a study of plant communities, data on eight functional traits of 30 herbs at 12 sites over 42 years were used to predict relative abundance (Shipley et al. 2006). Models could account for 94% of the observed variance. Similarly, Wootton (2005) found the neutral model fit relative abundance patterns in a rocky intertidal zone, but inclusion of species traits improved predictability of the results of experimental manipulations of the community.

A neutral biodiversity model based on network properties of rivers predicted fish community assemblages in the Mississippi-Missouri basin (Muneepeerakul et al. 2008). This application expands the approach from a two-dimensional approach to a dendritic network. This model was based on dispersal characteristics, so the results are similar to those for plants where consideration of functional attributes improved the model.

The broad view of the neutral model of biodiversity is that it represents one end of a continuum with niche-based models of communities forming the other (Gravel et al. 2006, May et al. 2007). If this continuum view is taken, falsification is not grounds for rejection of the neutral theory (Alonso et al. 2006), and could be indicative of the gradient. Alonso et al. (2006) mentioned that ideal gasses do not exist, just like completely neutral communities do not exist. Holt (2006) reviews some approaches that may reconcile niche-based models and predictions of neutral theory of biodiversity. He claims that neutrality can emerge from communities even when individual species do not meet the simple assumptions of neutrality. Hubbell (2006, p.1387) confirms this stance and states "The strategy behind neutral theory is to see how far one can get with the simplification of assuming ecological equivalence before introducing more complexity."

All this leaves us with a problem; what level of prediction is possible with the neutral theory of biogeography? Considerable empirical work is necessary before this question can be answered (Holyoak and Loreau 2006). Perhaps the controversy will die down and the neutral theory of biodiversity will become accepted as has neutral theory in population genetics (Hu et al. 2006). I have chosen to classify neutral "theory" as an important pattern of nature rather than a general theory, because of the uncertain answer to this question. However, some aspects of the approach using neutral theory form the basis of some of my proposed laws, thus the concept is important because it forces explicit definition of the null model.

What Is the Role of Natural History?

Natural history has a negative stereotype among some professional ecologists. One definition suggests it is the popular or amateur study of nature. Experimentally oriented ecologists have solidified the field of ecology, and this success has led some to think that natural history is descriptive and in no way experimental, so it should not be considered a true science. However, many branches of science are observational by necessity. Astronomy consists of observation of events that occurred longer ago the further away they are because of the time it takes light to reach the Earth. Many astronomical features cannot be the subject of experiments. Likewise, geology is concerned with many aspects of our planet that are historical, and observation must be used to establish pattern and process. I view natural history as the study of the specific contingencies of place including life histories and adaptations to other specific organisms. This knowledge of ecological history can strengthen the ability to make predictions and forms a vital part of the basis of ecology. Observation is the first step of the scientific method.

Natural history was defined by Peters (1991, after Hutchinson 1963, PANS Philadelphia 115:99–111) as an art, the goal of which is the personal and subjective development of the individual practitioner. If natural history is an art and not a science, it has little predictive ability. These definitions are dismissive of natural history, probably unnecessarily so.

Simberloff (2004, p. 787) claims that "Laws and models in community ecology are highly contingent, and their domain is usually local." If this is true, a good natural historian is the most likely to be able to predict the behavior of an individual system because of their correlative knowledge of local contingencies. If you have ever spent time in the field with a good natural historian, it is clear they understand a lot about how their ecological system of interest works. A good natural historian knows where and when plants and animals can be found, who eats what, and how climate influences organisms. Several factors such as flowering time, emergence from hibernation, arrival of migratory birds, and sequence of autumn leaf fall can be predicted within days to weeks. For example, the long records of flowering times have been used to demonstrate the ecological influence of global warming (Fitter and Fitter 2002), a use that was not conceived of when such records started to be kept. Evolutionary success of all species is a consequence of an ability to predict the natural world. Humans have taken this ability to extremes, and hominids were probably successful because they took a natural historical approach to predicting ecological features of their environment. We are probably genetically programmed to be natural historians. Furthermore, the beginning of prediction relies on detailed description of system dynamics and composition. Given the importance of contingencies in ecological systems, a careful accounting of local constraints is an essential part of describing patterns and can ultimately aid in ecological prediction. If Simberloff (2004) is correct that prediction is local and contingent, natural historical approaches could be the best way to approach predictive ecology.

Is There a General Theory of Ecological Scaling?

The problem of scale has concerned ecologists for many years. The general problem is that diversification means organisms operate on tremendously variable scales. Some organisms have a range of a millimeter or less, others migrate or disperse across continents. Simple animals can have a lifespan of a few weeks, whereas trees can live for many centuries. A small rotifer has fewer than 100 cells, and a blue whale has more than 10^{16} cells.

Allen and Starr (1982) made an influential attempt to explicitly synthesize the role of scale in ecology. Their ideas have received serious consideration, and their treatment remains highly cited in ecological literature. Allen and Starr (1982) make the point that many ecological systems are "middle number systems" that are neither so simple that it is easy to keep track of all individual components nor complex enough that statistics can be used to create laws (e.g., ideal gas laws that hold only when there are many molecules). Some aspects of ecological systems may be governed by such a large number of individuals (e.g., bacteria-dominated material fluxes in the oceans) that they qualify as large-number systems.

One difficulty with forming a predictive model of ecological scaling is the existence of emergent properties, properties that arise at one level as a result of simple dynamics at a lower level. A general theory of complex adaptive behavior of ecosystems is far from solidified. Research into general properties of complex systems (universal emergent properties) has yielded tantalizing insights but no general universal system (Kauffman 2000, Solé and Bascompte 2006). Marsh et al. (2006) suggest that emergence is a natural property of expansion of information of any system with multiple component parts governed by simple probabilistic rules. Exactly how this occurs is unknown, as is how to predict what emergent properties will arise in ecological systems.

Scale of processes and patterns can be found by analyzing characteristic length. Characteristic length is the scale under which controlling dynamics are most clearly observed (e.g., Henebry 1993). A characteristic length determination provides a subjective way to determine the scale at which an ecological pattern or process operates. Several techniques exist for teasing out the characteristic length. Emergent properties lead to multiple characteristic lengths as determined by these techniques (Habeeb et al. 2005). Thus, characteristic length analysis may remove some of the problems of dealing with scaling, but still can lead to the need to consider multiple scales. We do not know whether evolution leads species to key in on characteristic lengths of their environment.

One of the classical ways people have dealt with scale in biology and the way it is taught in introductory classes is the hierarchical organizational scheme: cell, organism, population, community, ecosystem, landscape, biosphere. Allen and Hoekstra (1992) make a clear case that this approach has distinct problems. Many of these problems arise from the law of diversification. A bacterial cell can be considered a single organism, but it is operating in spatial scales similar to individual cells in a complex animal. Individual cells in the animal are considered part of organs and the entire organism. The concept of causes being the most likely to occur at the level of organization below the entity of interest and the significance of a process being manifested in the level above is commonly used as a pattern to guide research.

An elegant way around the scaling problem in ecology is to determine scale-independent properties. Fractals have the property of scale invariance (fractals are defined as geometric patterns with scale-invariance [they look similar at all scales]), and fractal patterns are common in nature. Fractals will be discussed specifically in the next section. Scale-invariance is widespread in many nonequilibrium systems, including ecological systems (Solé and Bascompte 2006). However, scale invariance is not a universal feature of ecological systems. For example, there are minimum and maximum spatial and temporal scales that are relevant to individual organisms. Theoretical food web ecologists have proposed scale-independent properties, such as connectance or links per species, but many of these properties are conditional and specific to habitats and organisms (Montoya et al. 2006). Ecological networks do have unique patterns because they are built differently than other complex linked systems such as human social networks (Montoya et al. 2006). The problem is that these scale-independent properties are controversial and in many cases may not allow accurate predictions.

Another area in which scale independence has been proposed is landscape ecology, where self-similarity or fractal properties of habitats used by organisms are common. A potential problem with this approach is the law of specialization. Organisms must optimize their interactions on specific spatial and temporal scales or they will not be successful. Some organisms successfully deal with this dilemma by using different parts of their life cycle to operate at different spatial or temporal scales. For example, larvae of monarch butterflies (*Danaus plexippus*) have restricted spatial ranges but can eat low-quality food and still grow, whereas adults can migrate thousands of miles but need high-quality nectar to do so. *Daphnia* may reproduce every few weeks but can also produce resistant eggs (ephippia) able to survive in sediments for at least several decades (Hairston 1996).

There appears to be a negative relationship between organism size and population density that holds across many orders of magnitude; Schmid et al. (2000) found the relationship in two streams, one had 448 species, and the second had 260. This relationship has a large amount of variance (four orders of magnitude variation in population density at any body size), and the pattern only emerges when five to six orders of magnitude of sizes are considered.

Numerous allometric and metabolic patterns have been established that hold across a vast range of orders of magnitude of animal size. These patterns may indicate scale-independent dynamics that hold across many orders of magnitude (Brown et al. 2002). These scale-independent patterns may provide clues to fundamental laws of biology, although few noncontroversial laws have been established. The patterns are clear and indisputable (log-log trends, the ubiquity of power laws), but the actual slope of the relationships and the mechanism behind the slopes are not completely agreed on, and there is much unexplained scatter in many of these relationships.

Are Fractal Properties Related to Natural Selection?

The idea that nature has scaling properties that can be described by fractal (geometric patterns that are similar regardless of the scale of observation) mathematical relationships has generated considerable interest. Fractal scaling is common in many natural phenomena (Solé et al. 1999), and the possibility of using fractals to describe spatial ecological patterns is intriguing. Halley et al. (2004) identify just a few of the ways that fractal distributions may be generated that produce fractal patterns with potential ecological relevance: (1) inheritance—a fractal pattern may simply be a reflection of another underlying fractal (e.g., a fractal species distribution pattern may just reflect the fractal distribution of suitable habitat, (2) multi-scaled randomness—certain combinations of random processes, operating at different resolutions, generate statistical fractal output patterns (Solow 2005), (3) iterated mappings or successive branching rules (e.g., plant morphology), (4) diffusion-limited aggregation in which an object grows by accretion of randomly moving building blocks, (5) power-law dispersal with colonies established by relatively rare but sometimes long-distance migration (e.g., Levy Dust processes), (6) birth–death processes in which the birth process is random but the death process spatially aggregated or vice versa, and (7) self-organized criticality (patterns produced at the border between stability and chaos, they arise from the sum of interacting parts and do so in a nonlinear fashion) produces fractals in a variety of ways.

Allen and Hoekstra (1992) suggest that unpredictability in ecological systems occurs in regions where there are transitions between low- and high-fractal dimension. Organisms are evolutionarily successful if they are able to accurately predict their environment, so there should be selective pressure to take advantage of fractal properties if such properties are common enough in nature. Given the range of scales across which organisms must operate, it seems that the ability to operate across scales using simple relationships (e.g., moderately simple genetic programs), as occurs with fractal relationships, will lead to success. Indirect evidence that organisms operate across fractal scales according to fractal dimensions relates to the observation that activities of humans cause transitions between fractal dimensions. The disruptions of ecological systems may be related in part to the lowering of the predictability of the environment by changing the fractal dimension to which organisms are selected to respond. Part of the success of organisms may also be related to the ability to integrate across fractal dimensions. For example, the finer the resolution at which surfaces are observed, the more complex they become. Purposefully observing at larger scales will be useful for not getting lost in the details. Many animals can perceive a wide variety of environmental grain sizes.

Some potential problems in using fractal approaches to ecology (Halley et al. 2004) include: (1) power laws claimed when data for only a few orders of magnitude are available, (2) common fractal analyses such as "box counting" are overly sensitive to saturation of sampling and might display patterns of fractal dimension that are an artifact of sampling area, (3) linear regression is commonly used to estimate fractal dimension, but it can yield incorrect estimates when there is substantial scatter, (4) areas and borders of areas are commonly confused in fractal analyses by ecologists, (5) some patterns that are not inherently fractal will appear so, (6) it can be difficult to determine fractal properties of processes that are composed of two or more fractal properties, (7) definitions of fractal nature are imprecise, and (8) not all texture can be captured by fractal descriptions. Halley et al. (2004, p. 264) summarized these problems by stating "But just as it is clear that scale cannot be ignored, it is equally clear that few if any ecological phenomena are truly fractal."

QUESTIONS RELATED TO PATTERNS IN COMMUNITIES

The field of community ecology is an active area of ecological research yet includes some important, unresolved questions. First, the question of how interactions are distributed (e.g., are competitive interactions more prevalent than mutualistic) and the idea that indirect interactions may make ecological systems so complex that they defy prediction are considered. Second, the intermediate disturbance hypothesis and the relationship between diversity and stability are explored. Finally, I address a key issue of community ecology as related to management of ecological systems: Can we predict the success of invasive species?

What Is the Distribution of Interaction Types in Communities?

Four of the laws proposed in Chapter 1 have direct bearing on this question: *Species interact, All interaction signs possible, Diversity of interspecific interaction,* and *Variance of interspecific interaction.* Some community ecologists have taken the theoretical approach of investigating "competition" communities, and historical literature gives the impression that a view of the primacy of predation and competition in shaping communities is common.

More recently, the view that positive interactions can also constrain fundamental community structure has become more widespread (Travis et al. 2005). The laws proposed in Chapter 1 suggest a view of community structure that ignores mutualism, amensalism, commensalism, and neutralism should explicitly be recognized as simplified (Dodds 1997). If the laws above are accepted and all interaction types are possible, the question about distribution of interaction types becomes important because it is impossible to characterize a community correctly without taking into account all interaction types as well as their relative importance (Dodds and Nelson 2006). Stated differently, the null model is that all types of interaction are equally represented.

Few data are available on strengths and signs of species interactions in whole communities that are not biased because ecologists often use approaches based on their own background of taxonomic specialization (e.g., "communities" of birds or plants) or trophic specialization (e.g., "communities" of herbivores). The problem with these biases is that organisms that are closely related or those on the same trophic level may be more likely to compete relative to those at different trophic levels. Thus, a view that competition dominates community structure may be based on studies of groups of organisms that are mainly competitors.

If we take a completely unbiased sampling approach and randomly select two species from an area (e.g., a soil bacterium and a mouse with a burrow in the soil), what is the probable direct interaction between these two organisms? If we look at all such pairs of organisms or many such pairs selected without bias, what is the average interaction strength? Are interactions on balance positive or negative? Are they distributed normally? Are there long tails of the distribution so there are a few strong interactors (also referred to as keystone species)? These questions on community structure are currently unanswered.

Unidirectional direct species interactions measured in numerous studies of organisms from a single trophic level indicate that a range of interaction types are common in communities. The results of a meta-analysis of studies of species interaction indicates that many direct interaction strength distributions are normally distributed and a considerable number of interactions are either neutral (not significantly different from zero) or positive (Dodds and Nelson 2006). The distribution of interactions is an interesting result because species on the same trophic level need similar resources and should be expected to compete. Kaplan and Denno (2007) found a similar result when they reviewed 333 observations of phytophagous insects; competition was common, but they found asymmetric interactions, zero interactions, and even facilitative interactions in some studies.

The potential number of indirect interactions increases rapidly as the number of species considered increases (see Fig. 2). Conducting experiments on entire communities or even on species interactions among a substantial portion of the organisms in communities becomes difficult because so many interactions must be assessed. However, we do have more and more information on individual communities. Large-scale experiments in the field (e.g., Tilman et al. 1996) and controlled experimental systems (e.g., Naeem et al. 1994) are providing more information on species interactions among large groups of organisms. The large-scale field experiments may ultimately provide data that helps answer the question of distribution of interaction types in communities, at least under experimental conditions.

Meta-analysis suggests direct reciprocal interactions tend to be asymmetric; Vázquez et al. (2007) hypothesize that interaction strength depends in large part on relative abundance of the two species that are interacting, with the more abundant species having stronger interactions. Their meta-analysis supports the hypothesis of relative abundance driving interaction strength, but not in all cases. A much larger sample of communities would help discern general patterns.

Distribution of interaction types have been determined in only a few well-studied subsets of natural ecological communities (e.g., Seifert and Seifert 1976, Johansson and Keddy 1991, Paine 1992). The assumption that interspecific competition and predation are the primary drivers of community dynamics probably should not be made without a lot more data across ecosystems. Recent studies emphasizing facilitation and mutualism in the context of community structure support this newer view of community structure (Bertness and Callaway 1994, Bertness and Hacker 1994, Tirado and Pugnaire 2005).

Do Indirect Interactions Make Predictive Community Ecology Impossible?

Given the law that all species depend on other species and global biogeochemical linkages, most species on Earth are indirectly linked by a web of interactions. This web of dependence leads to the potential for irreducible complexity. Solé and Bascompte (2006) mention the potential ramifications of indirect interactions including the idea that reductionist views of ecosystems might not be possible with very strong indirect interactions and that evolutionary theory becomes difficult to expand to community scales. Yodzis (1988) analyzed a number of published food webs and noted that (1) responses of ecological communities to press experiments are unpredictable because of indirect interactions, (2) intraspecific interactions can be important, and (3) in some cases, indirect interactions are more important than direct interactions. However, communities could be highly compartmented, making it unlikely for indirect interaction chains to be propagated across many links.

Indirect interaction effects could be so strong and unpredictable that predictive community ecology of organisms embedded in a natural community is not possible (Dodds and Nelson 2006). Predicting responses of individuals in a community to changes in other species could be very difficult if weak interactions can transmit strong interactions through indirect interaction chains, interactions are nonlinear (e.g., density dependent), or interactions are highly variable over space and time. There is evidence that strong effects can be mediated by weak indirect effects in some webs (Berlow 1999) and it is difficult to quantify trait-mediated and density-mediated indirect interactions (Okuyama and Bolker 2007) to further complicate the problem. Interaction strength certainly varies over space and time.

Dodds and Nelson (2006) tested the general importance of indirect interactions using two alternate sets of assumptions and applied these assumptions to community interaction matrices from data published on 20 natural assemblages including plant and animal experiments from aquatic and terrestrial habitats with individual studies concentrated on single trophic levels. The first assumption applied to analyze the data was that interaction strengths were multiplicative. In this case, if interactions on average had a strength that was less than one, strength of the chain would diminish with length of the chain. An alternative assumption used to analyze the data was that interaction chains are no stronger than their weakest link. In both cases, interaction strength decreases as number of links tends to increase. It is possible that each individual species may be embedded in a web of interactions that have influence only over a finite number of links of indirect interaction chains radiating away from that species through the community. That is, indirect interactions attenuate in strength with number of links from each species, and many weak interactions tend to cancel each other, so there is an average number of species that are interacted with (perhaps an indirect interaction chain length <10) for each species. Navarrete and Berlow (2006) found that variability in indirect interaction can actually stabilize communities because of correlations of interaction strength with mortality and recruitment, further indication that unpredictable indirect interactions may not make community ecology impossible.

Tools are being developed to work with moderately complex systems (i.e., systems that are more complex than a few interacting individuals but not so complex that statistical relationships such as the ideal gas law can be applied). For example, Krause et al. (2003) used a social network method to analyze data from five complex empirical food webs and found evidence for compartmentalization in three of these. Communities can be stabilized if interactions are locally compartmented. Additional tools, such as neural network modeling and structural equations modeling, are providing researchers with methods to approach complex systems quantitatively.

If indirect interactions are strong and do not attenuate, general predictive community ecology (i.e., predicting abundance, distributions, and dynamics of many interacting species found in an area) may be difficult. In this case, only specific predictions could be made in heavily studied communities, and generalization of community interaction structure could be very difficult. As it stands, the question approached in this section remains unanswered.

Is the Intermediate Disturbance Hypothesis Useful?

There are several issues to consider when discussing whether the intermediate disturbance hypothesis can be used to predict a peaked relationship between diversity and disturbance. The first issue is the definition of disturbance. Pickett et al. (1989, p. 129) discussed the variety of definitions and provide the following, "Disturbance is a change in the minimal structure caused by a factor external of the level of interest." This is a broad definition, mainly because a disturbance for one species can be a benefit for another. For example, if a fire-tolerant grass has evolved to thrive in grasslands where fire is a regular occurrence, is fire a disturbance for this grass?

The second issue is how to scale disturbance for effect on communities. For ecosystems, we have no *a priori* method of scaling disturbance based on first principles. Intensity of disturbance cannot be mapped to intensity of effect on the community without prior calibration in an analogous ecosystem. Stated differently, we have no way to know what a disturbance is without prior experimentation (Huston 1994).

The third issue is that diversity has several properties, so more precise definition of the response variable is required. Research on phytoplankton in mesocosms indicates that richness and evenness can respond differentially to disturbance (Sommer 1995).

A maximum diversity occurs at the middle of a disturbance gradient in many habitats. These habitats include rolling rocks and algae in the marine intertidal zone (Sousa 1979), floods and invertebrates in streams (Townsend et al. 1997), variable dilution of phytoplankton cultures (depending how diversity is represented, Sommer 1995), tree falls in tropical forests (Molino and Sabatier 2001), and storms on coral reefs (Aronson and Precht 1995).

MacKey and Currie (2001) argued there is not a consistently peaked relationship between disturbance and diversity on the basis of 116 studies. They found that 10% to 20% of studies reviewed in the literature had a significant humped or peaked distribution. There was a range, because they grouped studies by response variable (richness, diversity, or evenness).

Why does the intermediate disturbance hypothesis apply at some times and not others? Disturbance effects are context-dependent. For example, terrestrial herbivore effects on diversity vary with soil fertility and precipitation (Olff and Ritchie 1998) and the presence of parasites can mediate disturbance responses (Morgan and Buckling 2004).

Discussion will be focused on streams since they are non-equilibrium systems, stream researchers have placed a great deal of attention on disturbance effects on biodiversity, and I am more familiar with this research. Floods are a common disturbance in streams and can be characterized as an evolutionary force over time (Poff 1992). The relationship between flood disturbance and invertebrate diversity in streams was investigated for 25 New Zealand streams (Death 2002). This study suggested that food availability (benthic algae) as influenced by floods and colonization dynamics were the most important indicators of species diversity. Research on prairie streams that considered both flood and drought supports the primacy of colonization in determining diversity and found no evidence for a humped distribution between habitat harshness and species diversity (Fritz and Dodds 2005). Similarly, McCabe and Gotelli (2000) found little support for the intermediate disturbance hypothesis explaining stream invertebrate responses to experimental disturbance regimes. Lake (2000) reviewed literature on flood disturbance and invertebrate diversity and found that disturbance effects depend upon scale of observation and that only some studies support the intermediate disturbance hypothesis.

Floods also can be used to illustrate difficulties in defining disturbances. Recurrence interval, maximum discharge, length of flood, or abruptness of rise all could influence biotic communities differently (Poff et al. 1997). Given the trade-off between lowered competition and food availability after floods as well as different effects on predators and competitors, it is not surprising that the intermediate disturbance hypothesis does not predict diversity in streams.

A relationship between diversity and disturbance is not unexpected given Pickett et al.'s (1989) definition of disturbance leading to a change in structure. The inability to predict the exact response across a disturbance gradient in a specific ecosystem without prior knowledge means that a humped relationship between diversity and disturbance cannot be assumed but could occur. Huston (1994) suggested humped shaped responses of diversity to disturbance should be most common in systems with high growth rates and intense competition. Roxburgh et al. (2004) argue there are several mechanisms of coexistence that can explain maximum diversity at intermediate disturbance and parsing out the individual mechanisms may ultimately lead to a broader-scope theory of diversity and disturbance. Application of the intermediate disturbance hypothesis requires knowledge of the systems of interest including life histories of the dominant organisms (Shea et al. 2004) and thus is predictive only within specific known ecosystems.

Is There a Relationship Between Diversity and Stability?

When modern theoretical ecology was developing, the concept that more diverse systems were more stable was common (MacArthur 1955, Hutchinson 1959). Stability-complexity debates have been ongoing for the last 50 years and are not yet resolved (e.g., Pimm 1991, Neutel et al. 2007). Urgency surrounding this issue has recently increased because of problems with invasive species and extirpations of native species. If diverse communities are more stable (in the sense of resistance to invasion by alien species), conserving biodiversity may also help keep noxious invasive species from becoming established.

A large part of the problem with establishing a link between diversity and stability is the variable definition of stability (Pimm 1991). We can return to the same site for many years in a row and find the same organisms in roughly the same relative abundances; this is probably the type of stability that most influenced early ecologists, who observed that "climax" communities do not change substantially over ecological time. Of course, this observation may be in part an illusion based on our scales of observation. Old-growth forests may seem constant, but long-term trends in climate change or disease make the forest community more dynamic. Ten thousand years is not a long ecological timeframe for a group of organisms that can live for centuries or even a thousand years.

At the simplest level, Elton (1958) noted the difficulty of maintaining monocultures of crops. The difficulty of maintaining monocultures of some species is empirically evident (and forms the basis of one of my laws, the requirement that species interact with other species, so single species cannot exist in isolation). However, considering more than a few species, the picture gets cloudy.

Robert May (1972, 1973) pointed out that matrices of interactions (all the direct interactions between all pairs of species in a community) are less stable as they become larger. The community matrices that May and others investigated assume modeled species interaction matrices are linear (assuming direct interactions are approximately linear when communities are close to equilibrium). This assumption was necessary for the mathematical techniques that were applied. Additional models with less restrictive assumptions have demonstrated that different assumptions can lead to more stable communities that are more complex and have defined how different types of stability (e.g., resilience and resistance) vary as a function of food web connectance and strength of interactions (Montoya et al. 2006). Allometric scaling among organisms (variety in size) in food webs could potentially stabilize diverse systems (Brose et al. 2006, Otto et al. 2007), and mutualistic relationships can also stabilize complex communities (Okuyama and Holland 2008), but these potential relationships lead to poor predictability across systems.

Empirical studies of plant communities also indicate increased stability in drought when diversity is greater (Frank and McNaughton 1991). In a long-term experiment on grassland plots, higher diversity plots had more stable production of aboveground biomass than lower diversity plots (Tilman et al. 2006); presumably, this would lead to a more stable food source for herbivores, and thus increased diversity of primary producers could enhance stability of food webs.

Prospects for a general relationship between stability and diversity are likely to remain unresolved. A positive correlation between stability and complexity is probably, at best, a weak pattern in nature. Predictability of the degree of stability as a function of species composition is currently not possible, although the idea that more diverse communities could be more resistant to invasion should not be ignored.

Are There Any General Rules That Can Be Used to Predict the Success of Invasive Species?

The problem of invasion by nonindigenous species has been recognized as an important issue in ecology over the last half century (Elton 1958). The problem is global in scope; species introductions cause substantial economic damage and harm to biodiversity worldwide (Vitousek et al. 1997). Thus, considerable effort has gone into predicting the probability that a habitat will be invaded by a nonindigenous species (Kolar and Lodge 2001). Some models for well-known species in specific areas have been honed to the point that invasion success can be predicted with 87% or better accuracy (Kolar and Lodge 2002). What are the potential predictive characteristics of species that invade successfully?

Moyle and Marchetti (2006) broke potential predictors of success into four areas: (1) commensal relationship of the species to humans, (2) life history of the invaders, (3) characteristics of the invasion site, and (4) invasion process. They then tested a series of predictions on fish invasions in California. Freshwater fishes are a good group to use because their taxonomy, distributions, and life history characteristics are well known and it is relatively easy to sample them reliably. Fish are confined to a comparatively limited area of the landscape, they are well sampled in many areas, and some species have been repeatedly and intentionally introduced into nonnative habitats. Results of this work suggest there are some potential general predictors of invasion success.

The first prediction is that species favored by humans have a high probability of becoming established. Species that are particularly successful have a mutualistic relationship with humans; both species benefit. There is no doubt mutualism aids in non-native establishment; crops, and livestock are the most widespread and conspicuous non-native species in many areas of the world. Similarly, fish species that humans desire because they are edible or have sport-fishing qualities (e.g., trout, carp, bass) have been introduced into fresh waters around the world. Species that are unlikely to succeed as invaders are not as likely to be chosen for introduction after a few failed attempts.

Aside from mutualistic species, Moyle and Marchetti (2006) mention as commensalistic species those that are transported and have no benefit to humans. Many invasive species that have been transported and become established have negative influences on humans (e.g., disease-causing organisms such as plague transmitted by rats), but others benefit with little negative effect on humans so could be considered commensalists.

The second area that may influence invasion success is life history including native range, broad physiological tolerance, genetic traits, r-life history (small size at first reproduction, rapid growth, and small adult body size), generalism, rapid dispersal, and novel traits or adaptations. The California fish data provide mixed or no support for large native range, r-life history, genetic traits, or generalism promoting invasion. Novel traits, such as the ability to feed on other non-native species, and reproductive strategies that maximize competitive success found some support in the California fish data. Broad physiological tolerance leading to invasion success found support because many aquarium species with limited tolerance for extreme environmental conditions were unable to establish.

The third area that may influence invasion success is characteristics of the invaded community. These include similarity to the habitat the non-native came from, human disturbance, and species richness. All three characteristics were supported by the California fish data, although contrary to theoretical expectations, more watersheds with more diverse fish assemblages also had more successful native invaders (see discussion on diversity stability issues).

The final areas related to the invasion process include transport, inoculation, establishment, dispersal, and integration. Transport rates of organisms have increased drastically with human activities. Increased movement of materials across and among continents and islands will move associated organisms, and those rates of movement will be greater than without human influence.

Kolar and Lodge (2001) analyzed quantitative studies of invasion success. They broke down the invasion process into transport, introduction, establishment, and spread. They found nine studies on introduction, and these studies supported the idea that release of a larger number of organisms or repeated releases increase the probability of establishment. They found that specific characteristics of species did not offer reliable broad predictability of success of invasion and that most of the successful life history attributes that lead to establishment were context-dependent. An exception to context-dependency was the observation that plants with a history of invading other habitats and those with vegetative reproduction are more likely to be successful invaders.

The somewhat tautological nature of this prediction (a successful invader is considered as such because it is so successful) is an example of contingency leading to more general prediction. We need to know details about a single species to predict whether it can invade a community. In conclusion, it is difficult to predict which specific species will invade successfully, but some patterns can be used to help predict relative probability of invasion.

QUESTIONS ABOUT ECOSYSTEMS AND COMMUNITY EFFECTS ON ECOSYSTEMS

The interface between community and ecosystem ecology has been fertile ground for ecological publication over the last decade. The effects of subsidies on food webs have garnered recent interest. The potential for trophic control of system primary production has generated discussion since the 1960s in part because of the fundamental ecological issues but also because food web manipulation has the potential for managing ecosystems. The relationship between biodiversity and ecosystem function is very important in a world where humanity is causing drastic declines in biodiversity. A number of recent reviews are considered that shed light on the predictive relationships between diversity and rates of ecosystem processes. Finally, I address the question of what controls nutrient retention in ecosystems because of the importance of understanding the effects of human-caused increases in the rates at which N and P enter the biosphere.

How Pervasive Are Food Web Subsidies?

Polis et al. (1997) argued that the landscape perspective dictates that subsidies will be a common property of food webs. This influential paper stimulated substantial research and a synthesis book on the topic (Huxel et al. 2004). Research in the area of food web subsidies is an approach that centers on the implications of a few of the fundamental laws that are proposed in this book: *Laws from physics and chemistry, System openness, Population heterogeneity.* The prior suggested laws, that resources and populations are distributed heterogeneously, that no ecosystem is closed, and that materials (and organisms with any random movement component) will diffuse from areas of high abundance to low abundance, dictate that food webs that abut areas of higher resources or greater population densities of organisms will be influenced by subsidies.

The most predictable subsidies are a result of those that go the assumed direction of material flows, with gravity (e.g., DeAngelis and Mulholland 2004). Gravitational transport of materials and diffusion theory can be combined to make general predictions about the direction of energy transport. In this sense, a theory of subsidies is possible.

However, much attention has been paid to examples where material flow is mediated by organisms and goes against gravity. Examples include movement of marine nutrients into terrestrial ecosystems on islands (Polis et al. 2004) and via migratory fish with special emphasis on salmon runs (Wilson et al. 2004). Many of these specific examples depend on contingencies derived from life histories of specific organisms. These contingencies are difficult to predict but can lead to significant material fluxes. There are a number of important examples. Vertical migration by zooplankton across mixing zones in lakes and marine environments can move materials much faster than molecular diffusion. Hippopotami and geese both graze in terrestrial environments and move nutrients into aquatic habitats where they rest and excrete. The rates of nutrient movement are substantially more rapid and more localized than abiotic transport processes. Benthic consumers that move through the water column of lakes (e.g., gizzard shad) move nutrients out of sediments more rapidly than advective transport, as do invertebrates that mix the sediments or pump water from the overlying waters to filter food or oxygenate their burrows. Beavers also move nutrients from riparian vegetation into streams and ponds more rapidly than would be predicted otherwise.

Perhaps the only generalization could be that organisms that move across habitats more rapidly than nutrients can be transported are likely to cause subsidies if they feed in nutrient enriched areas and move to those with lower nutrient availability. The problem with prediction is that organisms can increase heterogeneity (e.g., positive feedback, such as grazing lawns for ungulates) or decrease heterogeneity of resources by concentrating on rich patches and making them more resource poor. Still, the idea of subsidies provides a general pattern that should be considered by ecological researchers.

Can Top-Down Control Ultimately Control Primary Production?

If the law that the world is finite holds, something needs to limit abundance of primary producers. This limitation can come from below in the form of nutrients, light, abiotic disturbance, and, in terrestrial systems, water limitation. Limitation could also come from above (herbivory). The question then becomes, what limits the rate of herbivory and nutrient supply rate and what are their relative influences on primary producers? A simple view of propagation of interactions can lead to the theory that predator populations can control herbivores and the concept of the trophic cascade. Trophic cascade is based on a mathematical property of food chains; a negative effect on the trophic level below has a positive effect on the trophic level below that. This observation was made in 1880 by Lawrence Camarano (reprinted in 1994). Hairston et al. (1960) applied the reasoning to terrestrial ecosystems, and Brooks and Dodson (1965) applied the concept to fish effects on zooplankton grazing. The idea still generates discussion in ecological literature (e.g., Banse 2007).

The trophic cascade has been proposed as an important controller of lake productivity (Carpenter and Kitchell 1988). The presence or absence of large piscivorous fishes in lakes may alter production (e.g., Carpenter et al. 2001). One cross-study analysis suggests that the relative abundance of zooplanktivorous fish cannot be reliably correlated with algal biomass across unmanipulated lakes casting doubt on the ability to predict the occurrence and strength of trophic cascades (Currie et al. 1999). A second analysis of 54 pond or mesocosm manipulations of zooplanktivorous fish found a significant effect across studies. For the most part the effect was weak (30% change in phytoplankton in about two thirds of the studies) but it was strong in the remaining studies (Brett and Goldman 1996). Mazumder et al. (1990) found an effect of piscivorous fish on depth of summer surface water mixing (the epilimnion depth) as mediated by phytoplankton light attenuation in a more restricted group of lakes.

A detailed analysis of the strength of the cascade indicated that less than one-third of the variation in cascade strength across 114 studies could be predicted by system properties including metabolism, diversity, and taxonomy of herbivores and predators, and plant characteristics (Borer et al. 2005). Strength of the trophic cascade can vary with scale (Bell et al. 2003). Analyses of the trophic cascade in terrestrial systems indicated that 45 of 60 studies reported significant effects of manipulation of carnivores on plant response variables (Schmitz et al. 2000). An analysis across systems (terrestrial, marine, stream, lake) suggested top-down control of plant biomass was stronger in aquatic ecosystems than terrestrial and that the cascade most often decouples at the plant-herbivore interface (Shurin et al. 2002). A more extensive meta-analysis and associated modeling of 191 manipulative experiments verified a more consistent herbivore effect in aquatic habitats and suggested that compensation of herbivores, spatial and temporal heterogeneity, and nutrient recycling by herbivores may make prediction of top-down effects difficult (Gruner et al. 2008). Making things more complex, the food chain of a lake can have multiple stable states, in which changes lead to long-term shifts in the food chain. For example, when size structure of prey populations is not suitable, top down control can be ineffective unless size structure of the prey species (not the abundance of the predator species) is manipulated (Persson et al. 2007).

In a pointed criticism, Wetzel (2001) found two major problems with application of the concept of a trophic cascade in lakes. First, he noted that a variety of compensatory mechanisms emerge quickly after alterations in predation regimes (e.g., grazing-resistant morphologies of phytoplankton, increase in macrophytes following a decrease in phytoplankton). Second, Wetzel claimed the cascading influence cannot be sustained for long periods under natural conditions without disturbance because the cascades are only one potential route of several possible indirect interaction chains. If omnivory blurs food webs (Fagan 1997), cascades are more likely to operate in simple food webs such as the pelagic zone of lakes. Research on benthic algae demonstrates that grazing history does not influence the relative importance of top down control (Darcy-Hall and Hall 2008). Behavioral adaptations can also mediate trophic cascades (Beckerman et al. 1997, Schmitz et al. 1997, Schmitz 1998), and such behaviors can be difficult to predict. Invertebrate-mediated trophic cascades may be influenced by system productivity (Pringle et al. 2007), with maximum top-down effects at intermediate productivity (Schädler et al. 2003). The suggestion of top-down effects at intermediate productivity brings up arguments similar to those made about generality of prediction with respect to the intermediate disturbance hypothesis; what is "intermediate" productivity?

Despite criticism, trophic cascades have been described in a wide variety of systems, including rivers and streams, tropical forests, the open ocean (Pace et al. 1999), and coastal benthic habitats (Heck and Vallentine 2007). Top-down control can regulate production over short time periods in some habitats, but a variety of factors complicate prediction of when top-down control will occur. Some top-down effects are strong and indirect. Trophic cascades can be mediated by generalist predators with weak interactions with many herbivores (Moran and Hurd 1998). Foxes introduced to Aleutian Islands prey on seabirds, and this predation has cascading effects on plant communities. When the nutrient subsidies previously provided by the birds to the plants stopped, the islands' plant communities switched from grasses to shrub/forb dominated assemblages (Croll et al. 2005). In this case, alteration of food web subsidies cascaded to primary producers.

The debate in ecological literature is not whether trophic cascades can occur but rather how important top-down control is in specific systems. The trophic cascade is certainly a pattern ecologists should be aware of, particularly those working with simple systems where the food web has distinct trophic levels (where a food chain approximates the food web). Because each system lies on a gradient between bottom-up and top-down control of primary producers, it is difficult to predict where on this gradient a system lies without specific information on the habitat being considered. Thus, trophic cascades are an important pattern, but not as a predictive component of the general theory of community ecology presented in Chapter 2.

Is Biodiversity Predictably Linked to Ecosystem Function?

A recent area of research based on the idea that diverse assemblages may have more rapid rates of some key ecosystem processes has generated substantial interest. In the theoretical framework of this book I ask, can the theories of community structure and nutrient cycling be combined to make general predictions of the biotic effects on rates of ecosystem processes? Changes in biodiversity could have essential implications for ecosystem function (Tilman 1999), but the generality of this observation has generated controversy. The first step of the discussion is to define ecosystem function.

Ecosystem function can refer to any of the basal processes including primary production, detritivory, and nutrient cycling. Organisms may have conflicting influences on these various processes. For example, bison grazing on tallgrass prairie may limit rates of detritivory by lowering the rate at which detritus is produced (removing standing crop during the growing season of the grass). However, the same bison may increase the next season's production because in winter it eats dead grass (becomes a detritivore) and stimulates nutrient cycling and turnover by depositing feces and urine. Defining what specific ecosystem function is being influenced is important because of the conflicting roles that organisms play in various ecosystem processes.

At the most fundamental level, the hypothesis that biodiversity is related to ecosystem function is supported. Because one plant is more productive than no plants and one species of bacteria cycles nutrients more quickly than no bacteria, there is a positive relationship between biodiversity and ecosystem function. But according to the laws of *species interaction,* there will always be more than one species. The law of diversification suggests several species of each major functional group will exist. So now it can be asked, to what degree are species functionally redundant with respect to ecosystem functions?

There are complete nutrient cycles in even the simplest communities. For example, all major groups of bacteria can be isolated from deep groundwater habitats where microbes would be expected to be isolated, and perhaps less diverse (Sinclair and Ghiorse 1989). Thus, nutrient cycles will be complete regardless of minimal diversity. Stated differently, we know that multiple species of bacteria (syntrophy) are required to cycle organic compounds (break them down to CO_2 and methane) in anoxic environments (Fenchel and Finlay 1995), but an environment has yet to be described that does not contain this full complement. Biodiversity could influence rates of cycling in groundwater habitats dominated by microbes. The presence of grazing protozoa alters N cycling rates (Strauss and Dodds 1997) and can potentially influence rates of degradation of pollutants (Madsen et al. 1991). In groundwater sediments where pores are too small for protozoa to pass, but bacteria can, ecosystem function could be different (rates of cycling slower to respond to pulses) than in those where predation that increases efficiency of rates is present. We currently do not know enough about these simple communities to reliably predict how changes in biodiversity will influence ecosystem functions. Molecular techniques will allow microbial ecologists to approach such questions.

In a study using cultivable bacteria, Bell et al. (2005) examined constructed communities of up to 72 species with community respiration as the response variable indicating ecosystem function. They found that species richness continued to increase respiration, but not in a directly additive fashion. There was a saturating relationship with smaller increases in respiration as bacterial richness increased. Thus, functional redundancy was present, but richness had a positive relationship with the rate of ecosystem function.

A report commissioned by the Ecological Society of America provides a good synopsis of what can be predicted relative to biodiversity and ecosystem function (Hooper et al. 2005). The working group was certain that the following predictions (rephrased by me) could be made: (1) functional characteristics of species strongly influence ecosystems; (2) species invasions and extinctions caused by humans have altered ecosystem goods and services (ecosystem properties valued by humans) in ways that are impossible, difficult or costly to reverse; (3) effects of species loss or alterations in compositions are contingent and vary among ecosystems; (4) some ecosystem properties are decoupled from biodiversity because of functional redundancy, species with little influence on the ecosystem, or control by abiotic processes; and (5) as spatial and temporal variability increases, more species are needed to insure a stable supply of ecosystem goods and services. Interestingly, the report was certain that predictability was poor (predictions 3 and 4) in many cases. The authors expressed the opinion that poor ability to predict species effects on ecosystem goods and services should not be an excuse for humans to not worry about altering biodiversity through species introductions or extinctions.

Lake food webs are relatively simple, and research on these food webs can be used to assess the ability to link biodiversity to ecosystem properties. Alteration of a single trophic level in a food chain is the simplest type of biodiversity manipulation and is focused on an entire functional group. However, as discussed in the prior section on top-down-control, the ability to use trophic cascades in a predictive fashion, even in simple lake ecosystems, may be modest.

Stream researchers have attempted to link biodiversity to ecosystem function. Trophic cascades have been documented to occur in streams (Power 1990). However, these effects can be highly dependent on life histories of top predators (e.g., Huryn 1998). Leaf litter processing rates in streams may be greater when more species are present (Covich 1999). These effects could be limited to fairly species-poor streams such as those that occur on islands. Research from streams indicates that there can be links between biodiversity and ecosystem function, but the exact relationship is difficult to predict. Diversity of food sources (riparian vegetation) may maximize growth of detritiverous consumers in streams (Swan and Palmer 2006), suggesting that riparian tree diversity can relate to stream ecosystem production.

More diverse plant communities can lead to facilitative relationships that increase primary (Bertness and Leonard 1997) and secondary (Bertness et al. 1999) production. An example of these types of relationships is found in New England marsh plants (Bertness and Hacker 1994, Bertness and Leonard 1997). In this case, the grass *Juncus* lowers salinity and stimulates productivity of the marsh elder shrub (*Iva frutescens*). The marsh grass *Spartina* increases aeration of the sediments facilitating the other two plants. Together, all three plants decrease erosion and increase sedimentation, stabilizing the community. Viewed differently, facilitation may allow for more diverse communities (Tirado and Pugnaire 2005) that can influence ecosystem function. These results, though interesting, depend on the life histories of individual species and do not lead to general predictions, other than that the diversity of facilitators should be considered in many habitats, particularly those that are harsh.

Attempts have been made not only to link biodiversity to ecosystem process rates but also to the variability in these rates. France and Duffy (2006) used microcosms to test the effects of diversity and dispersal on final biomass of algae and invertebrates in sea grass communities. They found evidence that ecosystem structure was altered (variable amounts of edible algae and differences in biomass of primary consumers at the end of the experiment) and that increased diversity led to greater variability in these ecosystem properties. They did not directly test the effects on ecosystem function because they did not measure turnover of algae or invertebrates (production). The experiment is important because it implies that increases in diversity can decrease predictive ability.

The debate over diversity and ecosystem function has been most intense with regard to terrestrial plants. David Tilman et al. have published a number of articles claiming diversity in grasslands leads to higher production (Tilman and Downing 1994) and that loss of biodiversity threatens ecosystem function (Tilman et al. 1996). Some authors find that biodiversity effects on ecosystem function may be highly context-dependent (Wardle and Zackrisson 2005) or dependent on functional attributes of individual plants (Wright et al. 2006). Even more complexity enters the picture when multiple functions are considered; a greater diversity is required to support multiple ecosystem functions (Hector and Bagchi 2007).

Ability to predict potential maximum production based on grassland biodiversity may be useful for enhancing production rates of commodities. For example, grassland plants may provide a good source of carbon for biofuel production as well as high-quality livestock fodder. Certain polycultures may be more productive than simple monocultures more commonly used in modern agriculture (Tilman et al. 2006).

Bengtsson (1998) notes that types of species, the metric of biodiversity, and the ecosystem function of interest all need to be defined for specific cases before predictions can be made. Extrapolating from small-scale experiments linking diversity to some ecosystem property, to real systems occurring in nature could be difficult, primarily because of sampling effects (Cardinale et al. 2004). If multiple species respond to multiple stressors (e.g., what may occur with human effects such as global change), biodiversity and ecosystem function will both be influenced by the stressors (Vinebrooke et al. 2004), further confounding predictive ability.

A meta-analysis of 446 studies of biodiversity effects, 319 of which involved primary producer manipulations or measurements, indicated differential effects of diversity on ecosystem function (Balvanera et al. 2006). This analysis found variable effects of increasing biodiversity with strong positive effects on primary producer and primary consumer abundance as well as decomposer activity. Less consistent effects were found in characteristics such as community stability and secondary consumer abundance.

A meta-analysis of 111 field, laboratory, and greenhouse experiments indicated that loss of biodiversity can influence ecosystem function, but the exact species that is lost is the most important determinant of strength of effect (Cardinale et al. 2006). This analysis indicates that loss of the dominant or most productive species leads to the greatest change in ecosystem function and predicting in advance which species these are is difficult. The addition of trophic levels further complicates relationships between complexity and biodiversity (Duffy et al. 2007).

Research in aquatic and terrestrial systems confirms the difficulty of defining one simple overarching relationship between biodiversity and ecosystem function. A positive correlation between diversity and ecosystem functions is, however, a pattern to look for when attempting to link ecosystem properties to biotic characteristics. Relationships between biodiversity and ecosystem function could be particularly important across larger heterogeneous landscapes (Loreau et al. 2003).

What Controls Nutrient Retention in Ecosystems?

Nutrient retention is a concept that comes from simple budgets of nutrients. The difference between nutrient input and output is retention of the ecosystem. The time frame obviously needs to be defined because retention could be short- or long-term. The concept of nutrient retention has assumed new urgency as human influence on ecosystem N and P availability continues to increase.

Industrial processes by humans have doubled rates at which atmospheric N_2 is fixed in the biosphere. Similarly, phosphate is currently mined at rates equivalent to natural weathering (Vitousek et al. 1997). Agriculture to feed an expanding human population and appetite for foods from higher trophic levels (meat instead of vegetables) will only increase and expand the demand for N and P fertilizers. Continuous fertilization will require an understanding of nutrient retention to protect water quality and biotic integrity of aquatic ecosystems (Carpenter et al. 1998). The way nutrient cycles link, base productivity of ecosystems, and function within individual ecosystems all depend in part on how retentive ecosystems are of nutrients.

The concept that ecosystems can be saturated with nutrients and lose retention efficiency is central to ecosystem science. Odum (1969) claimed that nutrient retention became more efficient as succession proceeded. Within a few years, this idea was modified; extreme disturbance led to high initial nutrient export (Likens et al. 1970), but as biomass accumulated after the initial disturbance, retention increased to satisfy demand of organisms. Finally, when a successional climax was reached, nutrient retention decreased again (Vitousek and Reiners 1975). In systems such as urban areas, where forest is removed and fertilization continues, N retention stays low (Groffman et al. 2004).

Increases in atmospheric deposition of N have apparently led to decreases in ability to retain N in forested ecosystems (Aber et al. 1998). Little is known about how to reverse this saturation. Similarly, a central issue in control of N transport to downstream ecosystems is removal of N once it enters drainage networks. There is significant removal of N from river systems (Alexander et al. 2000), but it is not currently known where this removal occurs. A central issue to the location of removal is the functional effect of saturation of retention in aquatic environments (Bernot and Dodds 2005). Logically, there must be a maximum N retention that can occur before saturation processes occur. Uptake saturation has been characterized (O'Brien et al. 2007), but it is difficult to predict how this uptake saturation translates into long-term removal or storage.

CONCLUSIONS

Prediction is an elusive goal in any science; in ecology, the variance is great enough that the level of acceptable variance tolerated in prediction is substantially greater than in many other fields. Papers in ecology are accepted based on predicting only 40% of the variance, albeit with a high level of statistical significance. Though we would like to increase the amount of variance that is explained, we should be grateful for what we can explain. If scientists document that a disease has a 40% chance of killing each victim and establish with 95% certainty that there is a positive relationship between mortality and presence of the disease, that degree of certainty is generally sufficient to spend large amounts of money to prevent the disease. Physicists were willing to gamble that nuclear explosions from the first atomic test would not consume the earth, although they were not absolutely certain that would not occur. How certain is certain enough?

Many ecologists would disagree with my classifications of what areas of our scientific field are predictive enough to be considered theories. Some may think that what I call patterns deserve the status of a theory. I acknowledge that deciding how much prediction is necessary to elevate an area of research to theory status is approached in a subjective fashion. Hopefully, any controversy generated by my potentially idiosyncratic view at least stimulates discussion and ultimately leads to better science.

I have proposed some statements that allow general prediction in ecology in any specific situation. Combining multiple laws leads to incredible complexity, which is certainly a property of ecological systems. Still, theories can be built from those laws with predictive ability. These theories form the basis of the field of ecology. Having gone through the process of building theories from laws, I was surprised at the solidity of the framework behind predictive ecology. A predictive set of theories exists that allows understanding and management of our ecosystems. Typically, this prediction is reflected in patterns with high statistical significance but considerable residual variance (e.g., the relationship between nutrients and phytoplankton in lakes that is predicted by the theory of nutrient cycling).

General patterns such as those described in this chapter can guide ecologists to find less predictable properties of specific ecosystems. These patterns are not good for general prediction because they are too dependent on local contingencies. Still, these patterns can be useful. For example, the intermediate disturbance hypothesis does not always hold, but if an area is being managed to maintain natural diversity, the relationship between diversity and disturbance can be used to help maximize diversity in some cases, and managers should be aware that the relationship could be humped. These general patterns illustrate the idea that investigators can document patterns that are not broadly predictive but that can still be of practical use in specific situations.

My sense is that ecology is emerging as a predictive discipline. We now have computers that are powerful enough to run very complex models. We also have basic information on many habitats in electronic format, and we are developing protocols to combine disparate datasets. Tools such as spatial mapping software have vastly improved our ability to deal with complex spatial patterns, and multivariate statistical methods continue to improve. Simple features of ecological systems are becoming easier to combine to predict the behavior of complex interacting systems with biotic and abiotic components. With humans stretching the ability of global ecosystems to support us sustainably, the need for a predictive ecological science has never been greater. We are entering a "no-analog" ecological world (Williams and Jackson 2007), and general ecological principles that provide mechanisms and apply to such a world will be needed to predict its characteristics.

REFERENCES

Aber, J., W. McDowell, K. Nadelhoffer, A. Magill, G. Berntson, M. Kamakea, S. McNulty, W. Currie, L. Rustad, and I. Fernandez. 1998. Nitrogen saturation in temperate forest ecosystems. *BioScience* 48:921–34.

Alexander, R. B., R. A. Smith, and G. E. Schwarz. 2000. Effect of stream channel size on the delivery of nitrogen to the Gulf of Mexico. *Nature* 403:758–61.

Allen, A. P., J. F. Gillooly, V. M. Savage, and J. H. Brown. 2006. Kinetic effects of temperature on rates of genetic divergence and speciation. *PNAS* 103:9130–135.

Allen, A. P., and J. F. Gillooly. 2007. The mechanistic basis of the metabolic theory of ecology. *Oikos* 116:1073–077.

Allen, T. F. H., and T. W. Hoekstra. 1992. *Toward a unified ecology*. New York: Columbia Univ. Press.

Allen, T. F. H., and T. B. Starr. 1982. *Hierarchy*. Chicago: Univ. of Chicago Press.

Allen, T. F. H., R. B. O'Neill, and T. W. Hoekstra. 1984. Interlevel relations in ecological research and management: Some working principles from hierarchy theory. In *Readings in ecology,* ed. S. I. Dodson,

T. F. H. Allen, S. R. Carpenter, K. Elliot, A. R. Ives, R. L. Jeanne, J. F. Kitchell, N. E. Langston, and M. G. Turner, 393–412. New York: Oxford Univ. Press.

Alonso, D., R. S. Etienne, and A. J. McKane. 2006. The merits of neutral theory. *Trends Ecol. Evolution* 21:451–57.

Alonso, D., and M. Pascual. 2006. Comment on "A keystone mutualism drives pattern in a power function." *Science* 313:1739.

Andersson, A. F., and J. F. Banfield. 2008. Virus population dynamics and acquired virus resistance in natural microbial communities. *Science* 320:1047–050.

Aronson, R. B., and W. F. Precht. 1995. Landscape patterns of reef coral diversity: A test of the intermediate disturbance hypothesis. *J. Exp. Marine Biol. Ecol.* 192:1–14.

Azaele, S., S. Pigolotti, J. R. Benavar, and A. Maritan. 2006. Dynamical evolution of ecosystems. *Nature* 444:926–28.

Baguette, M. 2004. The classical metapopulation theory and the real, natural world: A critical appraisal. *Basic Appl. Ecol.* 5:213–24.

Balvanera, P., A. B. Pfisterer, N. Buchmann, J.-S. He, T. Nakashizuka, D. Raffaelli, and B. Schmid. 2006. Quantifying the evidence for biodiversity effects on ecosystem functioning and services. *Ecol. Letters* 9:1146–156.

Banse, K. 2007. Do we live in a largely top-down regulated world? *J. Bioscience* 32:791–96.

Beckerman, A. P., M. Uriarte, and O. J. Schmitz. 1997. Experimental evidence for a behavior-mediated trophic cascade in a terrestrial food chain. *Proceed. Natl. Acad. Sci.* 94:10735–0738.

Bell, G. 2001. Neutral macroecology. *Science* 293:2413–418.

Bell, T., W. E. Neill, and D. Schluter. 2003. The effect of temporal scale on the outcome of trophic cascade experiments. *Oecologia* 134: 578–86.

Bell, T., J. A. Newman, B. W. Silverman, S. L. Turner, and A. K. Lilley. 2005. The contribution of species richness and composition to bacteria services. *Nature* 436:1157–160.

Benda, L., N. L. Poff, D. Miller, T. Dunne, G. Reeves, G. Pess, and M. Pollock. 2004. The network dynamics hypothesis: How channel networks structure riverine habitats. *BioScience* 54:413–27.

Bengtsson, J. 1998. Which species? What kind of diversity? Which ecosystem function? Some problems in studies of relations between biodiversity and ecosystem function. *Appl. Soil Ecol.* 10:191–99.

Berlow, E. L. 1999. Strong effects of weak interactions in ecological communities. *Nature* 398:330–34.

Bernot, M. J., and W. K. Dodds. 2005. Nitrogen retention, removal, and saturation in lotic ecosystems. *Ecosystems* 8:442–53.

Berryman, A. A. 2003. On principles, laws and theory in population ecology. *Oikos* 103:695–701.

Bertness, M.D. and R. Callaway 1994. Positive interactions in communities. *Trends Ecol. Evolution* 9:191–93.

Bertness, M.D., and S. D. Hacker 1994. Physical stress and positive associations among marsh plants. *Am. Naturalist* 144:363–72.

Bertness, M. D., and G. H. Leonard. 1997. The role of positive interactions in communities: lessons from intertidal habitats. *Ecology* 78:1976–989.

Bertness, M.D., G. H. Leonard, J. M. Levine, P. R. Schmidt, and A. O. Ingraham 1999. Testing the relative contribution of positive and negative interactions in rocky intertidal communities. *Ecology* 80:2711–726.

Bodini, A. 1998. Representing ecosystem structure through signed digraphs. Model reconstruction, qualitative predictions and management: The case of a freshwater ecosystem. *Oikos* 83:93–106.

Bonner, J. T. 2006. *Why size matters: From bacteria to blue whales.* Princeton: Princeton Univ. Press.

Borer, E. T., E. W. Seabloom, J. Shurin, K. E. Anderson, C. A. Blanchette, B. Broitman, S. D. Cooper, and B. S. Halpern. 2005. What determines the strength of a trophic cascade? *Ecology* 86:528–37.

Bott, T. L., J. T. Brock, C. S. Dunn, R. J. Naiman, R. W. Ovink, and R. C. Peterson. 1985. Benthic community metabolism in four temperate stream systems: An inter-biome comparison and evaluation of the river continuum concept. *Hydrobiologia* 123:3–45.

Brett, M. T., and C. R. Goldman. 1996. A meta-analysis of the freshwater trophic cascade. *Proceed. Natl. Acad. Sci.* 93:7723–726.

Briand, F., and J. E. Cohen. 1987. Environmental correlates of food chain length. *Science* 238:956–60.

Brook, B. W., J. J. O'Grady, A. P. Chapman, M. A. Burgman, H. R. Akçakaya, and R. Frankham 2000. Predictive accuracy of population viability analysis in conservation biology. *Nature* 404:385–87.

Brook, B. W., and C. J. A. Bradshaw. 2006. Strength of evidence for density dependence in abundance time series of 1198 species. *Ecology* 87:1445–451.

Brooks, J. L., and S. I. Dodson. 1965. Predation, body size, and composition of plankton. *Science* 150:28–35.

Brose, U., A. Ostling, K. Harrison, and N. D. Martinez. 2004. Unified spatial scaling of species and their trophic interactions. *Nature* 428:167–70.

Brose, U., R. J. Williams, and N. D. Martinez. 2006. Allometric scaling enhances stability in complex food webs. *Ecol. Letters* 9:1228–236.

Brown, J. H., V. K. Gupta, B. Li, B. T. Milne, C. Restrepo, and G. B. West. 2002. The fractal nature of nature: Power laws, ecological complexity and biodiversity. *Philos. Trans. Royal Soc. London* B 357:619–26.

Brown, J. H. 1995. *Macroecology*. Chicago: Univ. of Chicago Press.

Brown, J. H., J. F. Gillooly, A. P. Allen, B. M. Savage, and G. B. West. 2004. Toward a metabolic theory of ecology. *Ecology* 85:1771–789.

Brown, J. H., J. G. Gillooly, G. B. West, and V. M. Savage. 2003. The next step in macroecology: From general empirical patterns to universal ecological laws. In *Macroecology: Concepts and consequences,* ed. T. M. Blackburn and K. J. Gaston, 408–23. Malden: Blackwell Publishing.

Buller, D. J. 2005. *Adapting minds: Evolutionary psychology and the persistent quest for human nature*. Cambridge: MIT press.

Camerano, L. 1994. On the equilibrium of living beings by means of reciprocal destruction. In *Frontiers in mathematical biology*, ed. S. A. Levin, 360–70. Berlin: Springer-Verlag.

Cardinale, B. J., A. R. Ives, and P. Inchausti. 2004. Effects of species diversity on primary productivity of ecosystems: Extending our spatial and temporal scales of inference. *Oikos* 104:437–50.

Cardinale, B. J., D. S. Srivastava, J. E. Duffy, J. P. Wright, A. L. Downing, M. Sankaran, and C. Jouseau. 2006. Effects of biodiversity on the functioning of trophic groups and ecosystems. *Nature* 443:989–92.

Carpenter, S. R., and J. F. Kitchell. 1988. Consumer control of lake productivity. *BioScience* 38:764–69.

Carpenter, S. R., N. F. Caraco, D. L. Correll, R. W. Howarth, A. N. Sharpley, and V. H. Smith. 1998. Nonpoint pollution of surface waters with phosphorus and nitrogen. *Ecological Applications* 8:559–68.

Carpenter, S. R., J. J. Cole, J. R. Hodgson, J. F. Kitchell, M. L. Pace, D. Bade, K. L. Cottingham, T. E. Essington, J. N. Houser, and D. E. Schindler. 2001. Trophic cascades, nutrients, and lake productivity: Whole-lake experiments. *Ecological Monographs* 71:163–86.

Carroll, J. W. 1994. *Laws of nature.* Cambridge: Cambridge Univ. Press.

Casjens, S. 1998. The diverse and dynamic structure of bacterial genomes. *Ann. Rev. Genet.* 32:339–77.

Chapelle, F. H., K. O'Neill, P. M. Bradley, B. A. Methé, S. A. Ciufo, R. L. Knobel, and D. R. Lovley. 2002. A hydrogen-based subsurface microbial community dominated by methanogens. *Nature* 415:312–15.

Chave, J. 2004. Neutral theory and community ecology. *Ecol. Letters* 7:241–53.

Clauset, A., C. Moore, and M. E. J. Newman. 2008. Hierarchical structure and the prediction of missing links in networks. *Nature* 98–101.

Cohen, J. E. 1995. Population growth and earth's human carrying capacity. *Science* 269:341–46.

Colwell, R. K., and D. W. Winkler. 1984. A null model for null models in biogeography. In *Ecological communities: conceptual issues and the evidence,* ed. D. R. Strong Jr., D. Simberloff, L. G. Abele, and A. B. Thistle, 344-59. Princeton: Princeton Univ. Press.

Colyvan, M., and L. R. Ginzburg. 2003. Laws of nature and laws of ecology. *Oikos* 101:649–53.

Connell, J. H. 1983. On the prevalence and relative importance of interspecific competition: evidence from field experiments. *Am. Naturalist* 122:661–96.

Cooper, G. J. 2003. *The science of the struggle for existence*. Cambridge: Cambridge Univ. Press.

Cottenie, K. 2005. Integrating environmental and spatial processes in ecological community dynamics. *Ecol. Letters* 8:1175–182.

Coulson, T., and H. C. J. Godfray. 2007. Single-species dynamics. In *Theoretical ecology: Principles and applications*, ed. R. M. May, and A. R. McLean, 17–34. Oxford, UK: Oxford Univ. Press.

Coulson, T., G. M. Mace, E. Hudson, H. Possingham. 2001. The use and abuse of population viability analysis. *Trends Ecol. Evolution* 16:219–21.

Covich, A. P. 1999. Leaf litter processing: The importance of species diversity in stream ecosystems. In *Biodiversity in benthic ecology*, ed. N. Friberg, and J. D. Carl, 15–20. NERI Technical Report No. 266. Denmark: National Environmental Research Institute.

Craine, J. M. 2005. Reconciling plant strategy theories of Grime and Tilman. *J. Ecol.* 93:1041–052.

Craine, J. M. 2007. Plant strategy theories: Replies to Grime and Tilman. *J. Ecol.* 95:235–40.

Croll, D. A., J. L. Maron, J. A. Estes, E. M. Danner, G. V. Byrd. 2005. Introduced predators transform subarctic islands from grassland to tundra. *Science* 307:1959–961.

Cuddington, K., and B. E. Biesner. 2005. Kuhnian paradigms lost: Embracing the pluralism of ecological theory. In *Ecological paradigms lost: Routes of theory change*, ed. K. Cuddington, and B. Beisner, 419–27. San Diego: Elsevier Academic Press.

Currie, D. J., P. Dilworth-Christie, and F. Chapleau. 1999. Assessing the strength of top-down influences on plankton abundance in unmanipulated lakes. *Can. J. Fisheries Aquatic Sci.* 56:427–36.

Cyr, H., and S. C. Walker. 2004. An illusion of mechanistic understanding. *Ecology* 85:1802–813.

Dagg, J. 2002. Unconventional bed mates: Gaia and the selfish gene. *Oikos* 96:182–86.

Dalsgaard, T., and B. Thamdrup. 2002. Factors controlling anaerobic ammonium oxidation with nitrite in marine sediments. *Appl. Environ. Microbiol.* 68:3802–808.

Darcy-Hall, T. L. and S. R. Hall. 2008. Linking limitation to species composition: Importance of inter- and intra-species variation in grazing resistance. *Oecologia* 155:797–808.

DeAngelis, D. L., and P. J. Mulholland. 2004. Dynamic consequences of allochthonous nutrient input to freshwater systems. In *Food webs at the landscape level*, ed. G. A. Polis, M. A. Power, and G. R. Huxel, 12–24. Chicago: Univ. of Chicago Press.

Death, R. G. 2002. Predicting invertebrate diversity from disturbance regimes in forest streams. *Oikos* 97:18–30.

Dodds, P. S., D. H. Rothman, and J. S. Weitz. 2001. Re-examination of the "3/4 law" of metabolism. *J. of Theoretical Biol.* 209:9–27.

Dodds, W. K. 1997. Interspecific interactions: Constructing a general, neutral model for interaction type. *Oikos* 78:377–83.

Dodds, W. K. 2002. *Freshwater ecology: Concepts and environmental applications*. San Diego: Academic Press.

Dodds, W. K. 2003. Misuse of inorganic N and soluble reactive P concentrations to indicate nutrient status of surface waters. *J. North Am. Benthological Soc.* 22:171–81.

Dodds, W. K. 2006. Nutrients and the "Dead Zone": Ecological stoichiometry and depressed dissolved oxygen in the northern Gulf of Mexico. *Frontiers Ecol. Environ.* 4:211–17.

Dodds, W. K. 2008. *Humanity's footprint: Momentum, impact, and our global environment.* New York: Univ. of Columbia Press.

Dodds, W. K., K. R. Johnson, and J. C. Priscu. 1989. Simultaneous nitrogen and phosphorus deficiency in natural phytoplankton assemblages: theory, empirical evidence, and implications for lake management. *Lake Reservoir Manage.* 5:21–26.

Dodds, W. K., and J. C. Priscu. 1990. A comparison of methods for assessment of nutrient deficiency of phytoplankton in a large oligotrophic lake. *Can. J. Fisheries Aquatic Sci.* 47:2328–338.

Dodds, W. K., A. J. López, W. B. Bowden, S. Gregory, N. B. Grimm, S. K. Hamilton, A. E. Hershey, E. Martí, W. B. McDowell, J. L. Meyer, D. Morrall, P. J. Mulholland, B. J. Peterson, J. L. Tank, H. M. Vallet, J. R. Webster, and W. Wollheim. 2002. N uptake as a function of concentration in streams. *J. North Am. Benthological Soc.* 21:206–20.

Dodds, W.K., and J.A. Nelson. 2006. Redefining the community: A species-based approach. *Oikos* 112:464–72.

Dodds, W.K., and J.J. Cole. 2007. Expanding the concept of trophic state in aquatic ecosystems: It's not just the autotrophs. *Aquatic Sci.* 69:427–39.

Dornelas, M., S.R. Connolly, and T.P. Hughes. 2006. Coral reef diversity refutes the neutral theory of biodiversity. *Nature* 440:80–82.

Douglas, M., and P.S. Lake 1994. Species richness of stream stones: an investigation of the mechanisms generating the species-area relationship. *Oikos* 69:387–96.

Driscoll, D.A. 2008. How to find a metapopulation. *Can. J. Zoology*. 85:1031–048.

Duffy, J.E., B.J. Cardinale, K.E. France, P.B. McIntyre, E. Thébault, and M. Loreau. 2007. The functional role of biodiversity in ecosystems: incorporating trophic complexity. *Ecol. Letters* 10:522–38.

Dunne, J.A., R.J. Williams, and N.D. Martinez. 2002. Food-web structure and network theory: The role of connectance and size. *PNAS* 99:12917–2922.

Egli, T. 1991. On multiple-nutrient-limited growth of microorganisms, with special reference to dual limitation by carbon and nitrogen substrates. *Antonie van Leeuwenhoek* 60:225–34.

Egli, T., and M. Zinn 2003. The concept of multiple-nutrient-limited growth of microorganisms and its application in biotechnological processes. *Biotechnol. Adv.* 22:35–43.

Elton, C.S. 1958. *The ecology of invasion by animals and plants*. New York: Wiley.

Enquist, B.J. 2002. Universal scaling in tree and vascular plant allometry: Toward a general quantitative theory linking plant form and function from cells to ecosystems. Tree Physiology 22:1045–064.

Ensign, S.H., and M.W. Doyle. 2006. Nutrient spiraling in streams and river networks. *J. Geophysical Res.* 111:G04009–G04022.

Evans-White, M. E., W. K. Dodds, and M. R. Whiles 2003. Ecosystem significance of crayfishes and stonerollers in a prairie stream: functional differences between co-occurring omnivores. *J. North Am. Benthological Soc.* 22:423–41.

Fagan, W. F. 1997. Omnivory as a stabilizing feature of natural communities. *Am. Naturalist* 150:554–67.

Fargione, J., C. S. Brown, and D. Tilman. 2003. Community assembly and invasion: An experimental test of neutral versus niche processes. *PNAS* 100:8916–920.

Fenchel, T., and B. J. Finlay. 1995. *Ecology and evolution in anoxic worlds*. Oxford, UK: Oxford Univ. Press.

Feynman, R. P. 1999. *The pleasure of finding things out: The best short works of Richard P. Feynman*. Cambridge, MA: Persus Publishing.

Finlay, B. J. 2002. Global dispersal of free-living microbial eukaryote species. *Science* 296:1061–063.

Fitter, A. H., and R. S. R. Fitter 2002. Rapid changes in flowering time in British plants. *Science* 296:1689–691.

France, K. E., and J. E. Duffy. 2006. Diversity and dispersal interactively affect predictability of ecosystem function. *Nature* 441:1139–143.

Frank, D. A., and S. J. McNaughton. 1991. Stability increases with diversity in plant communities: Empirical evidence from the 1988 Yellowstone drought. *Oikos* 62:360–62.

Fritz, K. M., and W. K. Dodds. 2005. Harshness: Characterization of intermittent stream habitat over space and time. *Marine Freshwater Res.* 56:13–23.

Giere, R. N. 1999. *Science without laws.* Chicago: The Univ. of Chicago Press.

Gilbert, B., W. F. Laurance, E. G. Leigh Jr., and H. E. M. Nascimento. 2006. Can neutral theory predict the responses of Amazonian tree communities to forest fragmentation? *Am. Naturalist* 168: 304–317.

Gillooly, J. F., J. H. Brown, G. B. West, V. M. Savage, and E. L. Charnov. 2001. Effects of size and temperature on metabolic rate. *Science* 293:2248–251.

Ginzburg, L. R., and M. Colyvan. 2004. *Ecological orbits: How planets move and populations grow*. New York: Oxford Univ. Press.

Ginzburg, L. R., and C. X. J. Jensen. 2004. Rules of thumb for judging ecological theories. *Trends in Ecol. and Evolution* 19:121–26.

Glazier, D. S. 2006. The 3/4-power law is not universal: evolution of isometric, ontogenetic metabolic scaling in pelagic animals. *BioScience* 56:325–32.

Gotelli, N. J., and C. M. Taylor. 1999. Testing metapopulation models with stream-fish assemblages. *Evolutionary Ecol. Res.* 1:835–45.

Gravel, D., C. D. Canham, M. Beaudet, and C. Messier. 2006. Reconciling niche and neutrality: the continuum hypothesis. *Ecol. Letters* 9:399–409.

Greenwood, J. L., and A.D. Rosemond. 2005. Periphyton response to long-term nutrient enrichment in a shaded headwater stream. *Can. J. Fisheries Aquatic Sci.* 62:2033–045.

Groffman, P. M., N. L. Law, K. T. Belt, L. E. Band, and G. T. Fisher 2004. Nitrogen fluxes and retention in urban watershed ecosystems. *Ecosystems* 7:393–403.

Gruner, D. S., J. E. Smith, E. W. Seabloom, S. A. Sandin, J. T. Ngai, H. Hillebrand, W. S. Harpole, J. J. Elser, E. E. Cleland, M. E. S. Bracken, E. T. Borer, and B. M. Bolker. 2008. A cross-system synthesis of consumer and nutrient resource control on producer biomass. *Ecol. Letters* 11:740–55.

Habeeb, R. L., J. Trebilco, S. Wotherspoon, and C. R. Johnson. 2005. Determining natural scales of ecological systems. *Ecological Monographs* 75:467–87.

Hairston, N. G. Sr., F. E. Smith, and L. B. Slobodkin. 1960. Community structure, population control, and competition. *Am. Naturalist* 94:421–25.

Hairston, N. G., Jr. 1996. Zooplankton egg banks as biotic reservoirs in changing environments. *Limnology Oceanography* 41:1087–092.

Hall, S. R., J. B. Shurin, S. Diehl, and R. M. Nisbet. 2007. Food quality, nutrient limitation of secondary production, and the strength of trophic cascades. *Oikos* 116:1128–143.

Halley, J. M., S. Hartley, A. S. Kallimanis, W. E. Kunin, J. J. Lennon and S. P. Sgardelis. 2004. Uses and abuses of fractal methodology in ecology. *Ecol. Letters* 7:254–71.

Hanski, I. 1999. *Metapopulation ecology*. Oxford: Oxford Univ. Press.

Hanski, I., and O. E. Gaggiotti. 2004. Metapopulation biology: Past, present, and future. In *Ecology, genetics, and evolution of metapopulations*, ed. I. Hanski, and O. E. Gaggiotti, 3–22. London: Elsevier.

Hanski, I., A. Moilanin, and M. Gyllenberg. 1996. Minimum viable metapopulation size. *Am. Naturalist* 147:527–41.

Hardin, G. 1960. The competitive exclusion principle. *Science* 131:1292–297.

Harrison, S. 1991. Local extinction in a metapopulation context: an empirical evaluation. *Biol. J. Linnean Soc.* 42:73–88.

Havens, K. E. 1992. Scale and structure in natural food webs. *Science* 257:1107–109.

Havens, K. E. 1993. Effect of scale on food web structure. *Science* 260:242–43.

Heck, K. L., and J. F. Valentine. 2007. The primacy of top-down effects in shallow benthic ecosystems. *Estuaries Coasts* 30:371–81.

Hector, A. and R. Bagchi. 2007. Biodiversity and ecosystem multifunctionality. *Nature* 448:188–91.

Henebry, G. M. 1993. Detecting change in grasslands using measures of spatial dependence with landsat TM data. *Remote Sensing Environ.* 46:223–34.

Hochachka, P. W., and G. N. Somero. 1984. *Biochemical adaptation*. Princeton: Princeton Univ. Press.

Holderegger, R., and H. H. Wagner. 2008. Landscape genetics. *BioScience* 58:199–207.

Holdridge, L. R. 1947. Determination of world plant formations from simple climatic data. *Science* 105:367–68.

Holt, R. E. 2006. Emergent neutrality. *Trends Ecol. Evolution* 21:531–33.

Holt, R. D., and T. H. Keitt. 2005. Species' borders: a unifying theme in ecology. *Oikos* 108:3–6.

Holt, R. D., J. H. Lawton, G. A. Polis, and N. D. Martinez. 1999. Trophic rank and the species-area relationship. *Ecology* 80:1495–504.

Holyoak, M., and M. Loreau. 2006. Reconciling empirical ecology with neutral community models. *Ecology* 87:1370–377.

Holyoak, M., M. A. Leibold, N. Mouquet, R. D. Holt and M. F. Hoopes. 2005. Metacommunities: A framework for large-scale community ecology. In *Metacommunities: Spatial dynamics and ecological communities*, ed. M. Holyoak, M. A. Leibold, and R. D. Holt, 1–31. Chicago: Univ. of Chicago Press.

Hooper, D. U., F. S. Chapin, III, J. J. Ewel, A. Hector, P. Inchausti, S. Lavorel, J. H. Lawton, D. Lodge, M. Loreau, S. Naeem, B. Schmid, H. Setälä, A. J. Symstad, J. Vandermeer, and D. A. Wardle. 2005. Effects of biodiversity on ecosystem functioning: A consensus of current knowledge. *Ecological Monographs* 75:3–35.

Horner-Devine, M. C., J. Lage, J. B. Hughes, and B. J. M. Hohannan. 2004. A taxa-area relationship for bacteria. *Nature* 432:750–53.

Howarth, R. W., and J. J. Cole. 1985. Molybdenum availability, nitrogen limitation and phytoplankton growth in natural waters. *Science* 229:653–55.

Hu, X., F. He, and S. P. Hubbell. 2006. Neutral theory in macroecology and population genetics. *Oikos* 113:548–56.

Hubbell, S. P. 2001. The unified neutral theory of biodiversity and biogeography. Princeton: Princeton Univ. Press.

Hubbell, S. P. 2006. Neutral theory and the evolution of ecological equivalence. *Ecology* 87:1387–398.

Hurlbert, S. H. 1984. Pseudoreplication and the design of ecological field experiments. *Ecological Monographs* 54:187–211.

Huryn, A. D. 1998. Ecosystem-level evidence for top-down and bottom-up control of production in a grassland stream system. *Oecologia* 115:173–83.

Huston, M. A. 1994. *Biological diversity: The coexistence of species on changing landscapes.* Cambridge: Cambridge Univ. Press.

Hutchens, E., S. Radajewski, M. G. Dumont, I. R. McDonald, and J. C. Murrell 2004. Analysis of methanotrophic bacteria in Movile Cave by stable isotope probing. *Environ. Microbiol.* 6:111–20.

Hutchinson, G. E. 1959. Homage to Santa Rosalia or why are there so many kinds of animals? *Am. Naturalist* 93:145–59.

Huxel, G. R., G. A. Polis, and R. D. Holt. 2004. At the frontier of the integration of food web ecology and landscape ecology. In *Food webs at the landscape level*, ed. G. A. Polis, M. E. Power, and G. R. Huxel, 434–51. Chicago: Univ. of Chicago Press.

Huxley, J. S. 1932. *Problems of relative growth*. New York: Dial Press.

Jassby, A. D., and T. Platt 1976. Mathematical formulation of the relationship between photosynthesis and light for phytoplankton. *Limnol. Oceanogr.* 21:540–47.

Jax, K. 2006. Ecological units: definitions and applications. *Q. Rev. Biol.* 81:237–58.

Jetz, W., C. Carbone, J. Fulford, and J. H. Brown. 2004. The scaling of animal space use. *Science* 306:266–68.

Jin, Q., and C. M. Bethke. 2003. A new rate law describing microbial respiration. *Appl. Environ. Microbiol.* 69:2340–348.

Joern, A., and S. Mole. 2005. The plant stress hypothesis and variable responses by Blue Grama Grass (*Boouteloua gracilis*) to water, mineral nitrogen, and insect herbivory. *J. Chem. Ecol.* 31:2069–2090.

Johansson, M. E., and Keddy, P. A. 1991. Intensity and asymmetry of competition between pairs of plants of different degrees of similarity: an experimental study on two guilds of wetland plants. *Oikos* 60:27–34.

Jørgensen, B. B. 1990. A thiosulfate shunt in the sulfur cycle of marine sediments. *Science* 249:152–54.

Jørgensen, S. E., and Svirezhev, Y. M. 2004. Towards a thermodynamic theory for ecological systems. Amsterdam: Elsevier.

Kaplan, I., and R. F. Denno. 2007. Interspecific interactions in phytophagous insects revisited: A quantitative assessment of competition theory. *Ecol. Letters* 10:977–94.

Kareiva, P., S. Watts, R. McDonald, and T. Boucher. 2007. Domesticated nature: shaping landscapes and ecosystems for human welfare. *Science* 316:1866–869.

Kauffman, S. A. 2000. *Investigations*. Oxford: Oxford Univ. Press.

Kauffman, S. A. 2008. *Reinventing the sacred: A new view of science, reason and religion*. New York: Basic Books.

Kerkhoff, A. J., and B. J. Enquist. 2007. The implications of scaling approaches for understanding resilience and reorganization in ecosystems. *BioScience* 57:489–99.

Kerr, J. T., H. M. Kharouba, and D. J. Currie. 2007. The macroecological contribution to global change solutions. *Science* 316:1581–584.

Kleiber, M. 1932. Body size and metabolism. *Hilgardia* 6:315–53.

Kneitel, J. M., and J. M. Chase. 2004. Trade-offs in community ecology: Linking spatial scales and species coexistence. *Ecol. Letters* 7:69–80.

Kolar, C. S., and D. M. Lodge. 2001. Progress in invasion biology: predicting invaders. *Trends Ecol. Evolution* 16:199–204.

Kolar, C. S., and D. M. Lodge. 2002. Ecological predictions and risk assessment for alien fishes in North America. *Science* 298:1233–236.

Kondratyev, K. Y., K. S. Losev, M. D. Ananicheva, and I. B. Chesnokaova. 2004. Stability of life on earth: principal subject of scientific research in the 21st century. Berlin: Springer-Verlag.

Kotliar, N. B., and J. A. Wiens. 1990. Multiple scales of patchiness and patch structure: a hierarchical framework for the study of heterogeneity. *Oikos* 59:253–60.

Koza, J. R. 1991. Genetic evolution and co-evolution of computer programs. In *Artificial life II, SFI studies in the sciences of complexity,* ed. C. G. Langton, J. D. Farmer, and S. Rasmussen, 603–28. Redwood City, CA: Addison-Wesley.

Krause, A. E., K. A. Frank, D. M. Mason, R. E. Ulanowicz, and W. W. Taylor. 2003. Compartments revealed in food-web structure. *Nature* 426:282–85.

Krebs, C. 1988. *The message of ecology*. New York: Harper Collins.

Krebs, C. J., S. Boutin, R. Boonstra, and A. R. E. Sinclair. 1993. Impact of food and predation of the snowshoe hare cycle. *Science* 269:1112–115.

Lake, P. S. 2000. Disturbance, patchiness, and diversity in streams. *J. North Am. Benthological Soc.* 19:573–92.

Lange, M. 2005. Ecological laws: what would they be and why would they matter? *Oikos* 110:294–403.

Lawton, J. H. 1999. Are there general laws in ecology? *Oikos* 84:177–92.

Latimer, A. M., J. A. Silander Jr., and R. M. Cowling. 2005. Neutral ecological theory reveals isolation and rapid speciation in a biodiversity hot spot. *Science* 309:1722–725.

Lee, G. F., and R. A. Jones. 1991. Effects of eutrophication on fisheries. *Rev. Aquatic Sci.* 5:287–305.

Leibold, M. A., M. Holyoak, N. Mouquet, P. Amarasekare, J. M. Chase, M. F. Hoopes, R. D. Holt, J. B. Shurin, R. Law, D. Tillman, M. Loreau, and A. Gonzalez. 2004. The metacommunity concept: A framework for multi-scale community ecology. *Ecol. Letters* 7: 601–13.

Levin, S. A. 1992. The problem of pattern and scale in ecology. *Ecology* 73:1943–967.

Levine, J. M., and C. M. D'Antonio. 1999. Elton revisited: A review of evidence linking diversity and invasibility. *Oikos* 87:15–26.

Levins, R. 1966. The strategy of model building in population biology. *Am. Scientist* 54:421–31.

Levins, R. 1969. Some demographic and genetic consequences of environmental heterogeneity for biological control. *Bull. Entomological Soc. Am.* 15:237–40.

Likens, G. E., F. H. Bormann, N. M. Johnson, D. W. Fisher, and R. S. Pierce. 1970. Effects of forest cutting and herbicide treatment on nutrient budgets in the Hubbard Brook watershed-ecosystem. *Ecological Monographs* 40:23–47.

Lindeman, R. L. 1942. The trophic-dynamic aspect of ecology. *Ecology* 23:399–417.

Link, J. 2002. Does food web theory work for marine ecosystems? *Marine Ecol. Progress Series* 230:1–9.

Lipton, P. 2005. Testing hypotheses: Prediction and prejudice. *Science* 307:219–21.

Lockwood, D. R. 2008. When logic fails ecology. *Q. Rev. Biol.* 83:57–64.

Loreau, M., N. Mouquet, and A. Gonzalez. 2003. Biodiversity as spatial insurance in heterogeneous landscapes. *PNAS* 100:12765–2770.

Lotka, A. J. 1925. *Elements of physical biology*. Baltimore: Williams & Wilkins Co.

Lowe, W. H., G. E. Likens, and M. E. Power. 2006. Linking scales in stream ecology. *BioScience* 56:591–597.

Lutz, R. A., and M. J. Kennish. 1993. Ecology of deep-sea hydrothermal vent communities: A review. *Rev. Geophysics* 31:211–242.

MacArthur, R. H. 1955. Fluctuations of animal populations and a measure of community stability. *Ecology* 36:533–536.

MacArthur, R. H. 1972. *Geographical ecology: Patterns in the distribution of species*. New York: Harper and Row.

MacArthur, R. H., and E. O. Wilson. 1967. *The theory of island biogeography*. Princeton: Princeton Univ. Press.

MacKey, R. L., and D. J. Currie. 2001. The diversity-disturbance relationship: Is it generally strong and peaked? *Ecology* 82:3479–492.

Madsen, E. L., J. L. Sinclair, and W. C. Ghiorse. 1991. In situ biodegradation: Microbiological patterns in a contaminated aquifer. *Science* 252:830–33.

Marsh, A. G., Y. J. Zeng, and J. Garcia-Frias. 2006. The expansion of information in ecological systems: Emergence as a quantifiable state. *Ecological Informatics* 1:107–16.

Martínez del Rio, C. 2008. Metabolic theory or metabolic models? *Trends Ecol. Evolution* 23:256–60.

Martinez, N. D. 1991. Artifacts or attributes? Effects of resolution on the Little Rock Lake food web. *Ecological Monographs* 61:367–92.

Martinez, N. D. 1993. Effect of scale on food web structure. *Science* 260:242–43.

Maurer, B. A. 1999. *Untangling ecological complexity: The macroscopic perspective*. Chicago: Univ. of Chicago Press.

May, R. M. 1972. Will large complex systems be stable? *Nature* 238:413–14.

May, R. M. 1973. *Stability and complexity in model ecosystems*. Princeton: Princeton Univ. Press.

May, R. M. 1976. Simple mathematical models with very complicated dynamics. *Nature* 261:459–67.

May, R. M., M. J. Crawley, and G. Sugihara. 2007. Communities: Patterns. In *Theoretical ecology: Principles and applications*, ed. R. M. May, and A. R. McLean, 111–31. Oxford, UK: Oxford Univ. Press.

Mayr, E. 2004. *What makes biology unique?* Cambridge: Cambridge Univ. Press.

Mazumder, A., W. D. Taylor, D. J. McQueen, and D. R. S. Lean. 1990. Effects of fish and plankton on lake temperature and mixing depth. *Science* 247:312–16.

McCabe, D. J., and N. J. Gotelli. 2000. Effects of disturbance frequency, intensity, and area on assemblages of stream macroinvertebrates. *Oecologia* 124:270–79.

McCann, K. S. 2000. The diversity-stability debate. *Nature* 405:228–33.

McGill, B. J. 2003. A test of the unified neutral theory of biodiversity. *Nature* 422:881–85.

McIntosh, R. P. 1985. *The background of ecology: Concept and theory.* Cambridge: Cambridge Univ. Press.

Miller, R. V. 1993. Genetic stability of genetically engineered microorganisms in the aquatic environment. In *Aquatic microbiology: An ecological approach*, ed. T. E. Ford, 483–511. Boston: Blackwell Publishing.

Miller, T. E., J. E. Burns, P. Munguia, E. L. Walters, J. M. Kneitel, P. M. Richards, N. Mouquet, and H. L. Buckley. 2005. A critical review of twenty years' use of the resource-ratio theory. *Am. Naturalist* 165:439–48.

Milne, B. T. 1998. Motivation and benefits of complex systems approaches in ecology. *Ecosystems* 1:449–56.

Molino, J.-F., and D. Sabatier. 2001. Tree diversity in tropical rain forests: A validation of the intermediate disturbance hypothesis. *Science* 294:1702–704.

Montgomery, D. R. 1999. Process domains and the river continuum. *J. Am. Water Resources Assoc.* 35:397–410.

Montoya, J. M., S. L. Pimm, and R. V. Solé. 2006. Ecological networks and their fragility. *Nature* 442:259–64.

Moran, M. D., and L. E. Hurd. 1998. A trophic cascade in a diverse arthropod community caused by a generalist arthropod predator. *Oecologia* 113:126–32.

Morgan, A. D., and A. Buckling. 2004. Parasites mediate the relationship between host diversity and disturbance frequency. *Ecol. Letters* 7:1029–034.

Moyle, P. B., and M. P. Marchetti. 2006. Predicting invasion success: Freshwater fishes in California as a model. *BioScience* 56:515–24.

Mulholland, P. J., A. M. Helton, G. C. Poole, R. O. Hall, Jr., S. K. Hamilton, B. J. Peterson, J. L. Tank, L. R. Ashkenas, L. W. Cooper, C. N. Dahm, W. K. Dodds, S. Findlay, S. V. Gregory, N. B. Grimm, S. L. Johnson, W. H. McDowell, J. L. Meyer, H. M. Valett, J. R. Webster, C. Arango, J. J. Beaulieu, M. J. Bernot, A. J. Burgin, C. Crenshaw, L. Johnson, B. R. Niederlehner, J. M. O'Brien, J. D. Potter, R. W. Sheibley, D. J. Sobota, and S. M. Thomas. 2008. Excess nitrate from agricultural and urban areas reduces dentrification efficiency in streams. *Nature* 452:202–06.

Muller-Landau, H. C., R. S. Condit, K. E. Harms, C. O. Marks, S. C. Thomas, S. Bunyavejchewin, G. Chuyong, L. Co, S. Davies, R. Foster, S. Gunatilleke, N. Gunatilleke, T. Hart, S. P. Hubbell, A. Itoh, A. R. Kassim, D. Kenfack, J. V. LaFrankie, D. Lagunzad, H. S. Lee, E. Losos, J.-R. Makana, T. Ohkubo, C. Samper, R. Sukumar, I.-F. Sun, N. Supardi, S. Tan, D. Thomas, J. Thompson, R. Valencia, M. I. Vallejo, G. Villa Munoz, T. Yamakura, J. K. Zimmerman, H. S. Dattaraja, S. Esufali, P. Hall, F. He, C. Hernandez, S. Kiratiprayoon, H. S. Suresh, C. Wills, and P. Ashton. 2006. Testing metabolic ecology theory for allometric scaling of tree size, growth and mortality in tropical forests. *Ecol. Letters* 9:575–88.

Muneepeerakul, R., E. Bertuzzao, H. J. Lynch, W. F. Fagan, A. Rinaldo and I. Rodriguez-Iturbe. 2008. Neutral metacommunity models predict fish diversity patterns in Mississippi-Missouri basin. *Nature* 453:220–23.

Murphy, P. G. 1975. Net primary productivity in tropical terrestrial ecosystems. In *Primary productivity of the biosphere*, ed. H. Lieth and R. H. Whittaker, 217–31. New York: Springer-Verlag.

Murray, B. M., Jr. 2000. Universal laws and predictive theory in ecology and evolution. *Oikos* 80:403–08.

Naeem, S. L., J. Thompson, S. P. Lawler, J. H. Lawton, and R. M. Woodfin. 1994. Declining biodiversity can alter the performance of ecosystems. *Nature* 368:734–37.

Navarrete, S. A., and E. L. Berlow. 2006. Variable interaction strengths stabilize marine community pattern. *Ecol. Letters* 9:526–36.

Neutel, A-M, J. A. P. Heesterbeek, J. van de Koppel, G. Hoenderboom, A. Vos, C. Kaldeway, F. Berendse, and P. C. De Ruiter. 2007. Reconciling complexity with stability in naturally assembling food webs. *Nature* 449:599–603.

Newbold, J. D., J. W. Elwood, R. V. O'Neill, and W. VanWinkle. 1981. Measuring nutrient spiraling in streams. *Can. J. Fish. Aquat. Sci.* 38:860–63.

Nowak, M.A. 2006. Five rules for the evolution of cooperation. *Science* 314:1560–563.

O'Brien, J. M., W. K. Dodds, K. C. Wilson, J. N. Murdock, and J. Eichmiller. 2007. The saturation of N cycling in Central Plains streams: ^{15}N experiments across a broad gradient of nitrate concentrations. *Biogeochemistry* 84:31–49.

O'Connor, M. P., S. J. Kemp, S. J. Agosta, F. Hansen, A. E. Sieg, B. P. Wallace, J. N. McNair and A. E. Dunham. 2007. Reconsidering the mechanistic basis of the metabolic theory of ecology. *Oikos* 116:1058–072.

Odum, E. P. 1959. *Fundamentals of ecology,* 2nd ed. Philadelphia: W. B. Saunders.

Odum, E. P. 1969. The strategy of ecosystem development. *Science* 164:262–70.

O'Hara, R. B. 2005. The anarchists guide to ecological theory. Or, we don't need no stinkin' laws. *Oikos* 110:390–93.

Okubo, A., and S. A. Levin. 2001. *Diffusion and ecological problems: Modern perspectives,* 2nd ed. New York: Springer-Verlag.

Okuyama, T., and B. M. Bolker. 2007. On quantitative measures of indirect interactions. *Ecol. Letters* 10:264–71.

Okuyama, T., and J. N. Holland. 2008. Network structural properties mediate the stability of mutualistic communities. *Ecol. Letters* 11:208–16.

Olff, H., and M. E. Ritchie. 1998. Effects of herbivores on grassland plant diversity. *TREE* 13:261–65.

Omerod, P. 2005. *Why most things fail: Evolution, extinction and economics*. New York: Pantheon Books.

Ostling, A., and J. Harte. 2003. A community-level fractal property produces power-law species-area relationships. *Oikos* 103:218–24.

Otto, S. B., B. C. Rall, and U. Brose. 2007. Allometric degree distributions facilitate food-web stability. *Nature* 450:1226–230.

Pace, M. L., J. J. Cole, S. R. Carpenter, and J. F. Kitchell. 1999. Trophic cascades revealed in diverse ecosystems. *Trends Ecol. Evolution* 14:483–88.

Paine, R. T. 1966. Food web complexity and species diversity. *Am. Naturalist* 100:65–75.

Paine, R. T. 1992. Food-web analysis through field measurement of per capita interaction strength. *Nature* 355:73–75.

Patrick, R. 1967. The effect of invasion rate, species pool, and size of area on the structure of the diatom community. *Proceed. Natl. Acad. Sci.* 58:1335–342.

Persson, L., P-A. Amundsen, A. M. De Roos, A. R. Knudsen, and R. Primicerio. 2007. Culling prey promotes predator recovery—alternative states in a whole-lake experiment. *Science* 316:1743–746.

Peters, R. H. 1983. *The ecological implications of body size*. Cambridge: Cambridge Univ. Press.

Peters, R. H. 1991. *A critique for ecology*. Cambridge: Cambridge Univ. Press.

Peterson, B. J., W. Wollheim, P. J. Mulholland, J. R. Webster, J. L. Meyer, J. L. Tank, N. B. Grimm, W. B. Bowden, H. M. Vallet, A. E. Hershey, W. B. McDowell, W. K. Dodds, S. K. Hamilton, S. Gregory, and D. J. D'Angelo. 2001. Control of nitrogen export from watersheds by headwater streams. *Science* 292:86–90.

Pickett, S. T. A., J. Kolasa, J. J. Armesto, and S. L. Collins. 1989. The ecological concept of disturbance and its expression at various hierarchical levels. *Oikos* 54:129–36.

Pickett, S. T. A., J. Kolasa, and C. G. Jones. 1994. *Ecological understanding: The theory of nature and the nature of theory*. San Diego: Academic Press.

Pickett, S. T. A., J. Kolasa, and C. G. Jones. 2007. *Ecological understanding: The theory of nature and the nature of theory*, 2nd ed. San Diego: Academic Press.

Pimm, S. L. 1991. *The balance of nature? Ecological issues in the conservation of species and communities.* Chicago: Univ. of Chicago Press.

Pimm, S. L. 2002. *Food webs*. Chicago: Univ. of Chicago Press.

Plattner, G.-K., F. Joos, and T. F. Stocker. 2002. Revision of the global carbon budget due to changing air-sea oxygen fluxes. *Global Biogeochemical Cycles* 16:1096, doi:10.1029/2001GB001.

Poff, N. L. 1992. Why disturbances can be predictable: A perspective on the definition of disturbance in streams. *J. North Am. Benthological Soc.* 11:86–92.

Poff, N. L. 1997. Landscape filters and species traits: Towards mechanistic understanding and prediction in stream ecology. *J. North Am. Benthological Soc.* 16:391–409.

Poff, N. L., J. D. Allan, M. B. Bain, J. R. Karr, K. L. Prestegaard, B. D. Richter, R. E. Sparks, and J. C. Strolmberg. 1997. The natural flow regime. A paradigm for river conservation and restoration. *BioScience* 47:769–84.

Polis, G. A., W. B. Anderson, and R. D. Holt. 1997. Toward an integration of landscape and food web ecology: the dynamics of spatially subsidized food webs. *Ann. Rev. Ecol. Systematics* 28:289–316.

Polis, G. A., F. Sánchez-Piñero, P. T. Stapp, W. B. Anderson, and M. D. Rose. 2004. Trophic flows from water to land: Marine input affects food webs of islands and coastal ecosystems worldwide. In *Food webs at the landscape level,* ed. G. A. Polis, M. A. Power, and G. R. Huxel, 200–16. Chicago: Univ. of Chicago Press.

Popper, K. 1968. *The logic of scientific discovery*. New York: Harper and Row.

Post, D. M., M. W. Doyle, J. L. Sabo, and J. C. Finlay. 2006. The problem of boundaries in defining ecosystems: A potential landmine for uniting geomorphology and ecology. *Geomorphology* 89:111–26.

Post, D. M., M. L. Pace, and N. G. Hairston Jr. 2000. Ecosystem size determines food-chain length in lakes. *Nature* 405:1047–049.

Post, W. M., C. C. Travis, and D. L. DeAngelis 1985. Mutualism, limited competition and positive feedback. In *The biology of mutualism,* ed. D. H. Boucher, 305–25. New York: Oxford Univ. Press.

Power, M. E. 1990. Effects of fish in river food webs. *Science* 250:811–14.

Power, M. E., D. Tilman, J. A. Estes, B. A. Menge, W. J. Bond, L. S. Mills, G. Daily, J. C. Castilla, J. Lubchenco, and R. T. Paine. 1996. Challenges in the quest for keystones. *BioScience* 46:610–20.

Pretzsch, H. 2002. A unified law of spatial allometry for woody and herbaceous plants. *Plant Biol.* 4:159–66.

Pretzsch, H. 2006. Species-specific allometric scaling under self-thinning: Evidence from long-term plots in forest stands. *Oecologia* 146:572–83.

Pringle, R. M., T. P. Young, D. L. Rubenstein, and D. J. McCauley. 2007. Herbivore-initiated interaction cascades and their modulation by productivity in an African savanna. *PNAS* 104:193–97.

Redfield, A. C. 1958. The biological control of chemical factors in the environment. *Am. Scientist* 46:205–21.

Reich, P. B., M. G. Tjoelker, J.-L. Machado, and J. Olkesyn 2006. Universal scaling of respiratory metabolism, size and nitrogen in plants. *Nature* 439:457–61.

Ricklefs, R. E. 2007. History and diversity: Exploration at the intersection of ecology and evolution. *Am. Naturalist supplement* 170:S56–S70.

Rose, M. R. 1987. *Quantitative ecological theory: An introduction to basic models*. London: Croom Helm.

Rosenzweig, M. L. 1968. Net primary productivity of terrestrial communities: Prediction from climatological data. *Am. Naturalist* 102:67–74.

Rosenzweig, M. L. 1995. *Species diversity in space and time*. Cambridge, MA: Cambridge Univ. Press.

Rosenzweig, M. L. 2003. *Win-win ecology: How the earth's species can survive in the midst of human enterprise*. Oxford: Oxford Univ. Press.

Ross-Gillespie, A., A. Gardner, S. A. West, and A. S. Griffin. 2007. Frequency dependence and cooperation: Theory and a test with bacteria. *Am. Naturalist* 170:331–42.

Roxburgh, S. H., K. Shea, and J. B. Wilson. 2004. The intermediate disturbance hypothesis: Patch dynamics and mechanisms of species coexistence. *Ecology* 85:359–71.

Ryberg, W. A., and J. M. Chase. 2007. Predator-dependent species-area relationships. *Am. Naturalist* 170:636–42.

Schädler, M., G. Jung, H. Auge and R. Brandl. 2003. Does the Fretwell-Oksanen model apply to invertebrates? *Oikos* 100:203–07.

Scheiner, S. M., and M. R. Willig. 2005. Developing unified theories in ecology as exemplified with diversity gradients. *Am. Naturalist* 166:458–69.

Scheiner, S. M., and M. R. Willig. 2007. A general theory of ecology. *Theoretical Ecology* DOI 10.1007/s12080-007-0002-0.

Schindler, D. W. 1998. Replication versus realism: The need for ecosystem-scale experiments. *Ecosystems* 1:323–34.

Schmid, P., E. M. Tokeshi, and J. M. Schmid-Araya. 2000. Relation between population density and body size in stream communities. *Science* 289:1558–560.

Schmidt-Neilsen, K. 1984. *Scaling: Why is animal size so important?* Cambridge, MA: Cambridge Univ. Press.

Schmitz, O. J. 1998. Direct and indirect effects of predation and predation risk in old-field interaction webs. *Am. Naturalist* 151:327–42.

Schmitz, O. J., A. P. Beckerman, and K. M. O'Brien. 1997. Behaviorally mediated trophic cascades: Effects of predation risk on food web interactions. *Ecology* 78:1388–399.

Schmitz, O. J., P. A. Hambäck, and A. P. Beckerman. 2000. Trophic cascades in terrestrial systems: A review of the effects of carnivore removals on plants. *Am. Naturalist* 155:141–53.

Schneider, E. D., and J. J. Kay. 1994. Life as a manifestation of the second law of thermodynamics. *Mathematical Computer Modeling* 19:25–48.

Schoener, T. W. 1983. Field experiments on interspecific competition. *Am. Naturalist* 122:240–85.

Schwartzman, D. W., and T. Volk. 1989. Biotic enhancement of weathering and the habitability of earth. *Nature* 340:457–60.

Seifert, R. P., and F. H. Seifert. 1976. A community matrix analysis of *Heliconia* insect communities. *Am. Naturalist* 110:461–82.

Shea, K., S. H. Roxburgh and E. S. J. Rauschert. 2004. Moving from pattern to process: Coexistence mechanisms under intermediate disturbance regimes. *Ecol. Letters* 7:491–508.

Shipley, B., D. Vile, and E. Garnier. 2006. From plant traits to plant communities: A statistical mechanistic approach to biodiversity. *Science* 314:812–14.

Shurin, J. B., E. T. Borer, E. W. Seabloom, K. Anderson, C. A. Blanchette, B. Broitman, S. D. Cooper, and B. S. Halpern. 2002. A cross-ecosystem comparison of the strength of trophic cascades. *Ecol. Letters* 5:785–91.

Sibly, R. M., D. Barker, J. Hone and M. Pagel. 2007. On the stability of mammals, birds, fish and insects. *Ecol. Letters* 10:970–76.

Simberloff, D. 2004. Community ecology: Is it time to move on? *Am. Naturalist* 6:787–99.

Simberloff, D. S. and E. O. Wilson. 1969. Experimental zoogeography of islands: The colonization of empty islands. *Ecology* 50:278–96.

Sinclair, J. L., and W. C. Ghiorse. 1989. Distribution of aerobic bacteria, protozoa, algae, and fungi in deep subsurface sediments. *Geomicrobiology J.* 7:15–31.

Solé, R. V., and J. Bascompte. 2006. Self-organization in complex ecosystems. Princeton: Princeton Univ. Press.

Solé, R. V., S. C. Manrubia, M. Benton, S. Kauffman and P. Bak. 1999. Criticality and scaling in evolutionary ecology. *Trends in Ecol. and Evolution* 14:156–60.

Solow, A. R. 2005. Power laws without complexity. *Ecol. Letters* 8:361–63.

Sommer, U. 1995. An experimental test of the intermediate disturbance hypothesis using cultures of marine phytoplankton. *Limnology and Oceanography* 40:1271–277.

Sousa, W. P. 1979. Disturbance in marine intertidal boulder fields: The nonequilibrium maintenance of species diversity. *Ecology* 60: 1225–239.

Stachowicz, J. J., R. B. Whitlatch, and R. W. Osman. 1999. Species diversity and invasion resistance in a marine ecosystem. *Science* 286:1577–578.

Stanford, P. K. 2006. *Exceeding our grasp: Science, history, and the problem of unconceived alternatives.* Oxford: Oxford Univ. Press.

Stanley, H. E., L. A. N. Amaral, S. B. Buldyrev, A. L. Goldberger, S. Havlin, H. Leschhorn, P. Maass, H. A. Makse, C.–K. Peng, M. A. Salinger, M. H. R. Stanley, and B. M. Viswanathan. 1996. Scaling and universality in animate and inanimate systems. *Physica* A 231:20–48.

Sterner, R. W., and J. J. Elser. 2002. *Ecological stoichiometry*. Princeton: Princeton Univ. Press.

Stoddard, P. K., and M. R. Markham. 2008. Signal cloaking by electric fish. *BioScience* 58:415–25.

Strauss, E. A., and W. K. Dodds. 1997. Influence of protozoa and nutrient availability on nitrification rates in subsurface sediments. *Microbial Ecol.* 34:155–65.

Strayer, D. L., M. E. Power, W. F. Fagan, S. T. A. Pickett, and J. Belnap 2003. A classification of ecological boundaries. *BioScience* 53:723–29.

Stream Solute Workshop. 1990. Concepts and methods for assessing solute dynamics in stream ecosystems. *J. North Am. Benthological Soc.* 9:95–119.

Swan, C. M., and M. A. Palmer. 2006. Composition of speciose leaf litter alters stream detritivore growth, feeding activity and leaf breakdown. *Oecologia* 147:469–78.

Thorp, J. H., M. C. Thoms, and M. D. Delong. 2006. The riverine ecosystem synthesis: biocomplexity in river networks across space and time. *River Res. Applications* 22:123–47.

Tilman, D. 1981. Tests of resource competition theory using four species of Lake Michigan algae. *Ecology* 62:802–15.

Tilman, D. 1982. *Resource competition and community structure*. Princeton: Princeton Univ. Press.

Tilman, D. 1999. The ecological consequences of changes in biodiversity: A search for general principles. *Ecology* 80:1455–474.

Tilman, D. 2007. Interspecific competition and multispecies coexistence. In *Theoretical ecology: Principles and applications,* ed. R. M. May and A. R. McLean, 84–97. Oxford, UK: Oxford Univ. Press.

Tilman, D., and J. A. Downing. 1994. Biodiversity and stability in grasslands. *Nature* 367:363–65.

Tilman, D., D. Wedin, and J. Knops. 1996. Productivity and sustainability influenced by biodiversity in grassland ecosystems. *Nature* 379:718–20.

Tilman, D., P. B. Reich, and J. M. H. Knops. 2006. Biodiversity and ecosystem stability in a decade-long grassland experiment. *Nature* 441:629–32.

Tirado, R., and F. I. Pugnaire. 2005. Community structure and positive interactions in constraining environments. *Oikos* 111:437–44.

Tonn, W. M., and J. J. Magnuson 1982. Patterns in the species composition and richness assemblages in northern Wisconsin lakes. *Ecology* 63:1149–166.

Townsend, C. R., M. R. Scarsbrook, and S. Dolédec. 1997. The intermediate disturbance hypothesis, refugia, and biodiversity in streams. *Limnol. Oceanogr.* 42:938–49.

Travis, J. M.J., R. W. Brooker, and C. Dytham. 2005. The interplay of positive and negative species interactions across an environmental gradient: insights from an individual-based simulation model. *Biol. Letters* 1:5–8.

Turchin P. 2001. Does population ecology have general laws? *Oikos* 94:17–26.

Turchin, P. 2003. *Complex population dynamics: A theoretical/empirical synthesis*. Princeton: Princeton Univ. Press.

Vandermeer, J., and I. Perfecto. 2006. A keystone mutualism drives pattern in a power function. *Science* 311:1000–002.

van der Meer, J. R. 2006. Environmental pollution selection of microbial degradation pathways. *Frontiers Ecol. Environ.* 4:35–42.

Vannote, R. L., G. W. Minshall, K. W. Cummins, J. R. Sedell, and C. E. Cushing. 1980. The river continuum concept. *Can. J. Fisheries Aquatic Sci.* 37:130–37.

Vázquez, D. P., C. J. Melian, N. M. Williams, N. Bluthgen, B. R. Krasnov, and R. Poulin. 2007. Species abundance and asymmetric interaction strength in ecological networks. *Oikos* 116: 1120–127.

Vinebrooke, R. D., K. L. Cottingham, J. Norberg, M. Scheffer, S. I. Dodson, S. C. Maberly, and U. Sommer. 2004. Impacts of multiple stressors on biodiversity and ecosystem functioning: The role of species co-tolerance. *Oikos* 104:451–57.

Vitousek, P. M., and R. W. Howarth. 1991. Nitrogen limitation on land and in the sea: How can it occur? *Biogeochemistry* 13:87–115.

Vitousek, P. M., H. A. Mooney, J. Lubchenco, and J. M. Melillo. 1997. Human domination of earth's ecosystems. *Science* 277:494–99.

Vitousek, P. M., and W. A. Reiners. 1975. Ecosystem succession and nutrient retention: a hypothesis. *BioScience* 25:376–81.

Volk, T. 1998. *Gaia's body: Toward a physiology of earth*. New York: Copernicus Springer-Verlag.

Vollenweider, R. A. 1976. Advances in defining critical loading levels for phosphorus in lake eutrophication. *Memorie dell'Istituto Italiano do Idrobiologia* 33:53–83.

Voltera, V. 1926. Fluctuations in the abundance of a species considered mathematically. *Nature* 2972:558–60.

Ward, J. V., and J. A. Stanford. 1983. Serial discontinuity concept of lotic ecosystems. In *Dynamics of lotic systems*, ed. T. D. Fontaine and S. M. Bartell, 29–42. Ann Arbor: Ann Arbor Science.

Wardle, D. A., and O. Zackrisson. 2005. Effects of species and functional group loss on island ecosystem properties. *Nature* 435:806–10.

Webster, J. R. 1975. Potassium and calcium dynamics in stream ecosystems on three southern Appalachian watersheds of contrasting vegetation. PhD dissertation. Athens, GA: Univ. of Georgia.

Webster, J. R., J. L. Tank, J. B. Wallace, J. L. Meyer, S. L. Eggert, T. P. Ehrman, B. R. Ward, B. L. Bennett, P. F. Wagner, and M. E. McTammany. 2000. Effects of litter exclusion and wood removal on phosphorus and nitrogen retention in a forest stream. *Verh Internat. Verein. Limnol.* 27:1337–340.

Webster's New Collegiate Dictionary. 1975. Springfield, MA: G. & C. Merriam Co.

West, G. B., J. H. Brown, and B. J. Enquist. 1999. The fourth dimension of life: Fractal geometry and allometric scaling of organisms. *Science* 284:1677–679.

West, G. B., J. H. Brown, and B. J. Enquist. 1997. A general model for the origin of allometric scaling laws in biology. *Science* 276:122–26.

Westermann, P. 1993. Wetland and swamp microbiology. In *Aquatic microbiology: An ecological approach,* ed. T. E. Ford, 215–38. Blackwell: Oxford.

Wetzel, R. G. 2001. *Limnology: Lake and river ecosystems,* 3rd ed. San Diego: Academic Press.

Wiegert, R. G., and C. E. Petersen. 1983. Energy transfer in insects. *Ann. Rev. Entomology* 28:455–86.

Wilkinson, D. M. 1999. Is Gaia really conventional ecology? *Oikos* 84:533–36.

Wilkinson, D. M. 2003. The fundamental processes in ecology: A thought experiment on extraterrestrial biospheres. *Biol. Rev.* 78:171–79.

Wilkinson, D. M. 2006. *Fundamental processes in ecology: An earth systems approach*. Oxford: Oxford Univ. Press.

Williams, J. W., and S. T. Jackson. 2007. Novel climates, no-analog communities, and ecological surprises. *Frontiers Ecol. Environ.* 5:475–82.

Wilson, D. S. 1992. Complex interactions in metacommunities, with implications for biodiversity and higher levels of selection. *Ecology* 73:1984–2000.

Wilson, M. F., S. M. Gende, and P. A. Bisson. 2004. Anadromous fishes as ecological links between ocean, fresh water and land. In *Food webs at the landscape level,* ed. G. A. Polis, M. A. Power, and G. R. Huxel, 12–24. Chicago: Univ. of Chicago Press.

With, K. A. 2004. Metapopulation dynamics: Perspectives from landscape ecology. In *Ecology, genetics, and evolution of metapopulations*, ed. I. Hanski and O. E. Gaggiotti, 23–44. San Diego: Academic Press.

Woodcock, S., C. J. van der Gast, T. Bell, M. Lunn, T. P. Curtis, I. M. Head, and W. T. Sloan. 2007. Neutral assembly of bacterial communities. *Microbial Ecol.* 62:171–80.

Wooton, R. J. 1990. *Ecology of teleost fishes*. London: Chapman and Hall.

Wootton, J. T. 2005. Field parameterization and experimental test of the neutral theory of biodiversity. *Nature* 433:309–12.

Wright, J. P., A. H. S. Naeem, C. Lehman, P. B. Reich, B. Schmid, and D. Tilman. 2006. Conventional functional classification schemes underestimate the relationship with ecosystem functioning. *Ecol. Letters* 9:111–20.

Yedid, G., and G. Bell. 2002. Macroevolution simulated with autonomously replicating computer programs. *Nature* 420:810–12.

Yodzis, P. 1988. The indeterminacy of ecological interactions as perceived through perturbation experiments. *Ecology* 69:508–15.

Yoon, I., R. J. Williams, E. Levine, S. Yoon, J. A. Dunne, and N. D. Martinez. 2004. Webs on the Web (WoW): 3D visualization of ecological networks on the WWW for collaborative research and education. *Proceed. IS&T/SPIE Symp. Electronic Imaging Visualization Data Anal.* 5295:124–32.

Yoshinaga, T., A. Hagiwara, and K. Tsukamtol. 2001. Why do rotifer populations present a typical sigmoid growth curve? *Hydrobiologia* 446:99–105.

Zavaleta, E. S., and K. B. Hulvey. 2004. Realistic species loss disproportionately reduces grassland resistance to biological invaders. *Science* 306:1175–177.

INDEX

Composition	:	Aptara Corp., Inc.
Text	:	11/15 Granjon
Display	:	Granjon
Printer and Binder	:	Maple Vail